Lecture Notes in Civil Engineering

Volume 19

Lecture Notes in Civil Engineering (LNCE) publishes the latest developments in Civil Engineering—quickly, informally and in top quality. Though original research reported in proceedings and post-proceedings represents the core of LNCE, edited volumes of exceptionally high quality and interest may also be considered for publication. Volumes published in LNCE embrace all aspects and subfields of, as well as new challenges in, Civil Engineering. Topics in the series include:

- Construction and Structural Mechanics
- Building Materials
- Concrete, Steel and Timber Structures
- Geotechnical Engineering
- Earthquake Engineering
- Coastal Engineering
- Hydraulics, Hydrology and Water Resources Engineering
- Environmental Engineering and Sustainability
- Structural Health and Monitoring
- Surveying and Geographical Information Systems
- Heating, Ventilation and Air Conditioning (HVAC)
- Transportation and Traffic
- Risk Analysis
- Safety and Security

To submit a proposal or request further information, please contact the appropriate Springer Editor:

- Mr. Pierpaolo Riva at pierpaolo.riva@springer.com (Europe and Americas);
- Ms. Swati Meherishi at swati.meherishi@springer.com (India);
- Ms. Li Shen at li.shen@springer.com (China);
- Dr. Loyola D'Silva at loyola.dsilva@springer.com (Southeast Asia and Australia/NZ).

More information about this series at http://www.springer.com/series/15087

Mokhtar Awang · Mohamed Hasnain Isa
Editors

The Advances in Civil Engineering Materials

Selected Papers of the ICACE 2018 held in Batu Ferringhi, Penang Malaysia on 9th–10th May 2018

Springer

Editors
Mokhtar Awang
Department of Mechanical Engineering
Universiti Teknologi Petronas
Seri Iskandar, Perak, Malaysia

Mohamed Hasnain Isa
Department of Civil and Environmental Engineering
Universiti Teknologi Petronas
Seri Iskandar, Perak, Malaysia

ISSN 2366-2557 ISSN 2366-2565 (electronic)
Lecture Notes in Civil Engineering
ISBN 978-981-13-2510-6 ISBN 978 981-13-2511-3 (eBook)
https://doi.org/10.1007/978-981-13-2511-3

Library of Congress Control Number: 2018954020

This Springer imprint is published by the registered company Springer Nature Singapore Pte Ltd.
The registered company address is: 152 Beach Road, #21-01/04 Gateway East, Singapore 189721, Singapore

Contents

Investigation of Housing Preferences in View of the Housing Market Dynamics in Doha Metropolitan

Hatem Ibrahim[1(✉)], Ashraf Salama[2], Florian Wiedmann[2], Reem Awwaad[1], and Bassma Aboukalloub[1]

[1] Department of Architecture and Urban Planning, College of Engineering, Qatar University, Doha, Qatar
hatem_ibrahim@qu.edu.qa
[2] Department of Architecture, University of Strathclyde Glasgow, Glasgow, UK

Abstract. Since the 2000s, Doha has gone through enormous economic growth and urbanization processes. The massive migration inflows have suggested particular housing preferences that are driven by the demographic and lifestyle diversities of the multicultural population. Housing preferences of the multicultural population are defined in terms of housing location, cost, and typology. The paper aims to study housing preferences in view of the housing market dynamics in order to recommend future actions for housing planning and expansion. It focuses on the period from 2000 to 2017 where rapid urban development took place in Doha metropolitan considering cultural, sporting, economic, real estate, and political incidents. The methodological approach includes: questionnaire survey of housing preferences, and survey of the perceptions of housing market experts. The paper defines the major shifts in the housing market based on: the sporting events, construction boom, and real-estate boom. It will discuss solutions to reduce the gap between housing supply and demand. Also, it will conclude actions for future housing expansion based on the recommendations by key market experts towards a sustainable local economic growth.

Keywords: Housing preferences · Multicultural population · Housing market
Doha metropolitan

1 Introduction

The rapid urbanization and the massive inflows of the multicultural population have posed significant questions to the housing market in Doha metropolitan. Today, Doha's multicultural population is made-up of four main cultural groups: Qataris, Arabs, Asians, and Westerners. In this paper, Qataris are referred to as nationals; while Arabs, Asians, and Westerners are referred to as migrants. In 2016, migrants have made-up 88% of Qatar's total population (Snoj 2017). This reflects the intensity of multi-cultures where the needs of each cultural group are different in Doha metropolitan.

M. Awang and M. H. Isa (eds.), *The Advances in Civil Engineering Materials*,
Lecture Notes in Civil Engineering 19, https://doi.org/10.1007/978-981-13-2511-3_1

Studying the housing preferences will aid in developing sustainable neighborhoods (Petkar and Macwan 2013). Housing preferences are known to be related to demographic factors on how nationals and migrants are residing in the city (Haybatollahi et al. 2015). The demographic structure of the nationals is different than that of migrants. Housing preferences can be defined as the expression of the quantity and quality of housing features (Lee et al. 2008; Hunt 2010; Ouwehand and Doff 2011). In this paper, housing preferences are defined by three factors: location, cost, and typology.

2 Research Methodology

2.1 Research Tools

The paper uses a quantitative and qualitative research methods through the following tools:

1. Survey of housing preferences that is conducted through a questionnaire, which is distributed to 388 residents in Doha for them to study the demographic pattern of the multicultural population.
2. Survey of experts' perceptions in the housing market is conducted through face-to-face interviews from local authorities, including real estate developers, architects, urban planners, and academics.

2.2 Study Area and Period

Doha metropolitan is composed of Doha municipality, parts of Al-Rayyan, Um Slal, and Al-Daayen municipalities (Fig. 1). Throughout the paper, Doha metropolitan is referred to as merely Doha. Neighborhoods in Doha are divided into: downtown, suburban, and waterfront. Downtown neighborhoods represent the old center of Doha up to the C-Ring Road. Suburban neighborhoods include all urban areas around the D, E, and F-Ring roads. Waterfront neighborhoods include the coastal line of Doha, except the part attached to the downtown neighborhoods.

The study focuses on the period of rapid housing dynamics, starting from year 2000 to 2017. The highly relevant incidents during this period have affected the development of housing typologies:

- Sporting events: hosting the Asian Games in 2006 and winning the World Cup bid in 2010.
- Construction boom: the establishment of artificial island of the Pearl Qatar in 2011 and the announcement of Qatar Integrated Railway Project in 2012.
- Real-estate boom: approval of property ownership rule for migrants, established in 2004.

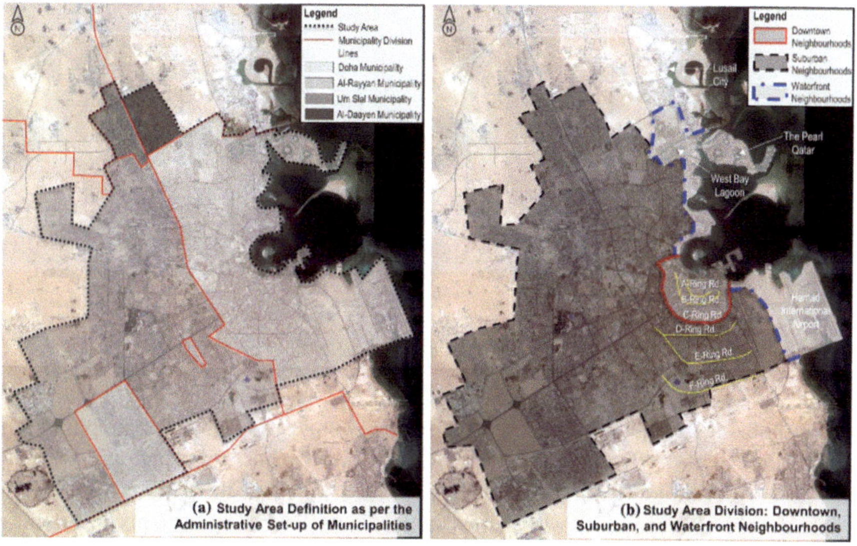

Fig. 1. The study area of Doha metropolitan and neighborhood divisions. *Source* Ministry of Development Planning and Statistics (2013), Ministry of Development Planning and Statistics (2015) and Ministry of Municipality and Environment (2015) "edited by the authors"

3 Research Background

During the 2000s, the housing market in Doha witnessed the biggest influence of the population's diverse demographics (Fig. 2a). Land use planning was based on residential land use distribution (Jaidah and Bourennane 2009; Shandas et al. 2017). Traditionally, the downtown was the most sought-after location for housing in Doha (Eddisford and Carter 2017). Gradually as the city began to move away from traditional locations into areas around the main ring roads, these locations grew in importance for housing developments (Salama and Wiedmann 2013). A preference to live near the workplace, school, and lifestyle facilities has guided housing location preferences (Scheiner and Kasper 2003; Dawkins and Koebel 2009; Pisman et al. 2011). Today, the waterfront has become a sought-after location for high-end housing targeting high-income groups such as the West Bay Lagoon, Lusail City, and The Pearl Qatar (Property Market Overview 2017).

According to Colliers International (2014), 70% of Doha's population are of low-income groups (Fig. 2b). Since 70% of Doha's population earn a monthly salary form 5000 QR to 19,999 QR (low-income group), they are not able to find their appropriate housing based on their preferences. This can be due to the rapid population growth, which has created a shortage of affordable housing supply. Since the existing supply of housing targets middle to high-income groups, it can be stated that the housing needs of Doha's population are not met. This reflects the strong need for affordable housing. This also implies the influence of affordability levels on housing preferences, which makes the population compromise their housing needs in order to save costs. Although the government is seeking to address this issue, affordable housing is expected to remain in short supply.

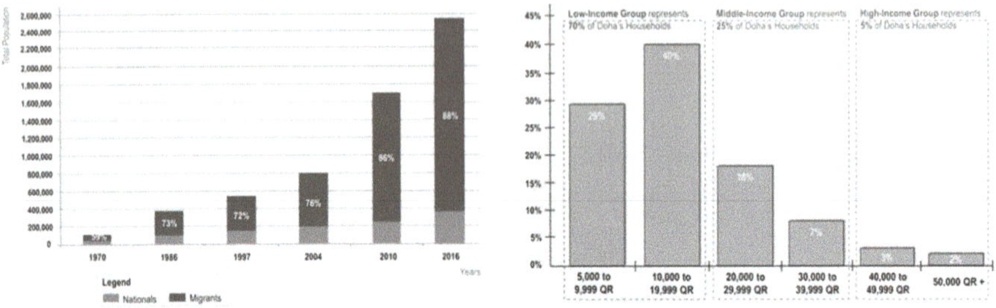

(a) The multicultural population in Doha metropolitan (source: Bel-Air, 2014; Snoj, 2017)

(b) Dominant income group of Doha's multicultural population (source: Colliers International, 2014)

Fig. 2. Characteristics of the multicultural population in Doha metropolitan. *Source* Developed by the authors

The housing market has mirrored the urban changes in Doha. The population has steadily moved away from the high-density downtown into suburban and waterfront locations. According to Colliers International (2013), most of the nationals live in villas in suburban neighborhoods, while most of migrants live in apartments in downtown neighborhoods. This has shaped the housing dynamics in Doha where apartments are dominating the market (Wiedmann et al. 2012). Therefore, it can be concluded that housing demand in terms of typologies is clearly defined by nationality.

4 Housing Preferences and the Market Dynamics

The analysis of the questionnaires reflects almost the actual demographic and housing conditions of Doha. The existing demographic composition of the population is clearly reflected in which 92% of total respondents are migrants, which reflects that the majority of Qatar's population are migrants. In terms of the cultural mix, Asians are the largest cultural group in Doha, which is reflected in the questionnaire where 41% of the respondents are Asians (Fig. 3a). In terms of gender, male respondents are more than female respondents, (Fig. 3b), which is confirmed by the Bel-Air (2014), where men have outnumbered women by almost 1–5.

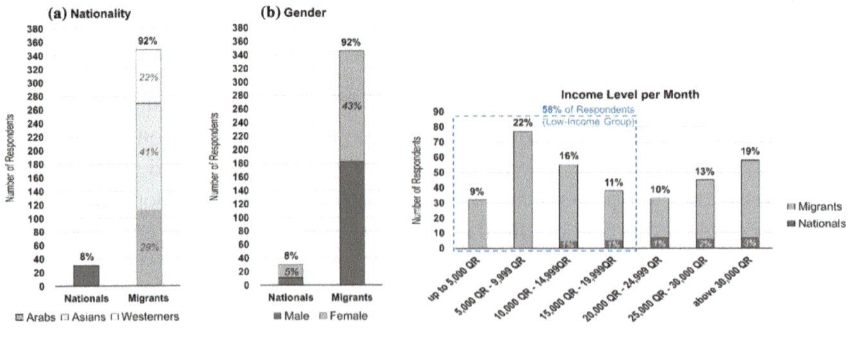

Fig. 3. Demographic data of respondents. *Source* Based on the questionnaire survey

4.1 Housing Location

It is concluded that 53% of the respondents live in suburban neighborhoods, where single-family housing (villa typology) is popular. Whereas 33% live in downtown neighborhoods, and 14% live in waterfront neighborhoods where multi-family housing (apartment typology) is popular (Fig. 4a). In terms of housing location preferences, 39% prefer living in waterfront neighborhoods, in which 38.5% of them are migrants, which reflects their strong preference for waterfront neighborhoods. This can be due to the real estate law that allows housing ownership for migrants. However, nationals prefer suburban neighborhoods in which 7% prefer suburban neighborhoods (Fig. 4b).

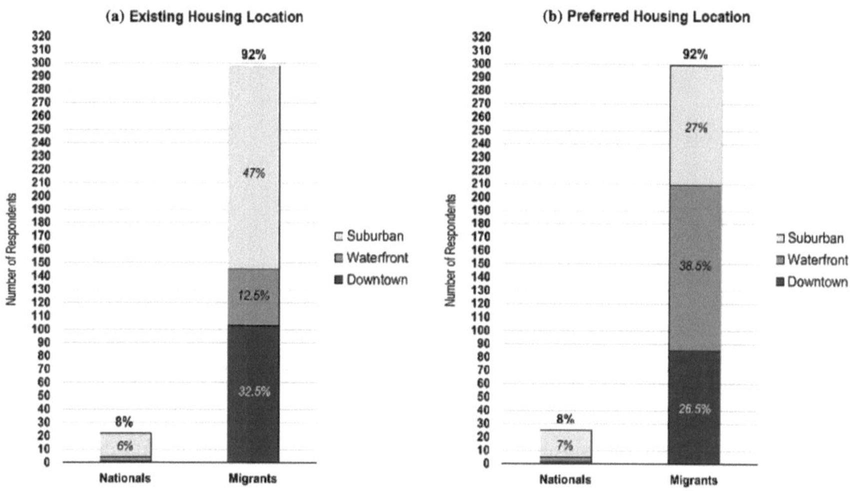

Fig. 4. Housing location preferences. *Source* Based on the questionnaire survey

Experts have agreed that the governmental vision is the main driver for urban and population growth in Doha. Major incidents took place resulting in construction and real estate booms in the city. However, housing developments have been impacted by a number of factors related to planning and zoning, accessibility, and equity. Zoning and building regulations have the greatest impact on housing developments where most of the existing housing designs are not committed to the local culture, climate, and context.

4.2 Housing Cost

In terms of housing costs analysis, 90.5% of respondents live in rental housing where migrants are the majority (90%), (Fig. 5a) and would still prefer to rent (63%), (Fig. 5b). This can be due to the income level and short-term stay in Doha. In reference to the nationals housing which the government commonly supports, 7.5% own their houses and 8% would prefer to own. The housing market has been affected by the multiplicity of income level diversities. However, this is not the case with nationals where almost all of them are house owners (7.5%). Therefore, housing rentals play a considerable role in adjusting housing preferences in Doha where the ideal housing environment can be compromised in order to save costs, especially for migrants (67%), (Fig. 5c).

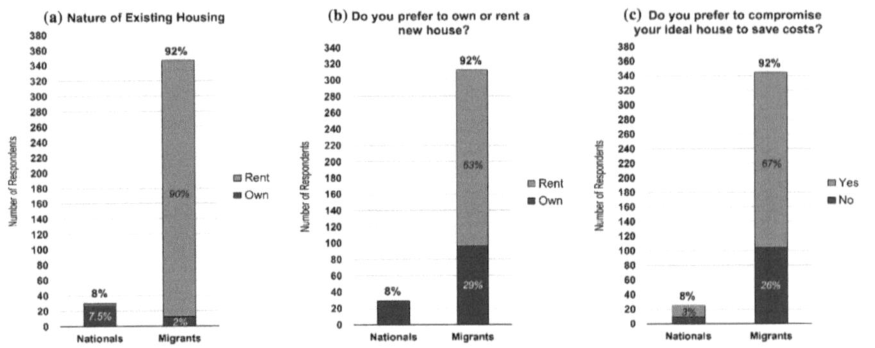

Fig. 5. Housing cost preferences. *Source* Based on the questionnaire survey

Typically, the income level drives housing preferences in Doha. Nationals tend to value community residence and spaciousness (housing location and typology). On the other hand, migrants tend to value costs and affordable living (housing cost). Therefore, affordability is among the significant reasons for favouring apartments especially for single migrant workers.

Experts have focused on the growth of population in Doha and referred to the construction boom in 2004 that increased the migration flows. It can be concluded that the housing market of Doha lacks the standard levels of affordability. Low-income migrants struggle to find their appropriate housing. The improved housing regulations and policies can address this need through regulating housing prices and quality.

4.3 Housing Typology

In terms of housing typology analysis, 64% of the migrants live in apartments (Fig. 6a). Yet, only 27.5% of migrants prefer apartments, and the majority preferred villas (64.5%). However, an interesting fact was concluded for nationals who all live in villas (8%), but 2.5% prefer apartments (Fig. 6b). This can reflect the need of nationals for housing diversity. This implies that typology preferences of the multicultural population are not addressed since the existing market supply serves the work-oriented nature of the migrants. It can be concluded that the diverse housing typologies, which address family living, based on the culture of each population group, are a common preference in Doha.

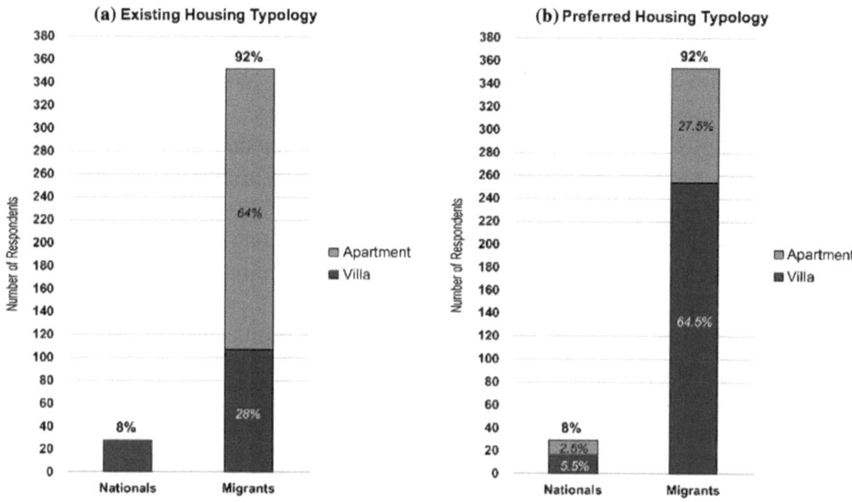

Fig. 6. Housing typology preferences. *Source* Based on the questionnaire survey

It is concluded that housing typology preferences in Doha reflect the cultural origins of the population. The villa typology is dominant in suburban neighbourhoods, which is aligned with the preferences of nationals. It is the most preferred typology where spacious rooms and large living areas are provided to accommodate the family-oriented activities and frequent gatherings. However, apartments are the most supplied typology in the market, especially in suburban and waterfront neighbourhoods. The needs and values of the multicultural population should be addressed through a preference-driven supply of housing typologies with proper costs in preferable locations.

In terms of real estate conditions, experts have agreed on the importance of the contribution of the government and the large and small-scale developers in the local housing market. A common agreement is concluded that income groups guide the housing market in Doha. The supply by real estate developers targets particular income groups, in which particular housing areas and typologies are being developed. Some suggestions have emphasized on the role of the government to support real estate developers through regulations and incentives. At the end, the target is to have a housing market that is accessible to all cultural and income groups.

5 Conclusion and Recommendations

According to the analysis, it is realized that among the most prevailing factors affecting urban growth of Doha and its peripheries are the decisions that are posed by the government. A set of recommended actions were developed which summarized the solid outcomes of the paper (Fig. 7).

- Adopt a decentralized approach for housing expansion towards other organized suburban centers to the north of Doha.

- Include housing policies that consider mix of typologies and densities to reduce the effect of social segregation in Doha.
- Revitalize downtown neighborhoods towards lifestyle housing communities that satisfy both nationals and migrants' housing preferences.
- Provide social housing to address issues related to housing inequality and affordability. The government should contribute with real estate developers towards affordable housing provision in Doha.
- Allow cooperation between government (regulations) and the real estate developers (housing developments) to reduce the pressure on the city's infrastructure through considering nationals and migrants needs. In terms of local economic growth, this would be an important step towards a more balanced housing market.
- Encourage developers to build affordable housing in Doha through governmental incentives.
- Allow housing ownership for migrants towards sustainable economic growth.
- Plan and design walkable neighborhoods and housing communities.

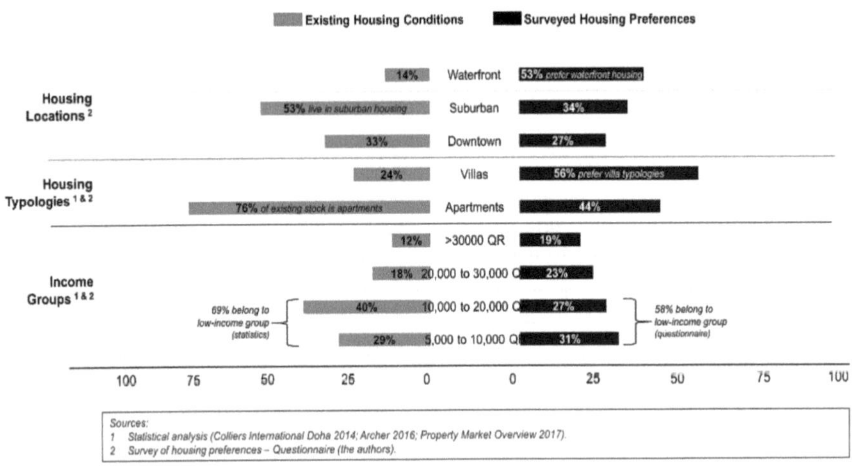

Fig. 7. Comparative summary of the existing housing conditions and the surveyed housing preferences in Doha metropolitan. *Source* Based on the statistical analysis and housing preferences analysis

Most of the experts agreed that the low-income groups would struggle to find a suitable housing unit; they share units as a reflection of their need for affordable housing. In terms of the future market outlook, experts suggest to provide opportunities for migrants to own houses as they are considered final investors for housing in Doha. Additionally, legislation should be cooperated with the design of housing to reduce the pressure on the city's infrastructures and government. Suggestions were introduced to improve the policies, which respect to the local culture, climate, and context. This includes more opportunities for migrants to own their housing in Doha towards a livable housing market.

Acknowledgements. This paper was made possible by NPRP grant number [NPRP 07-960-5-135] from the Qatar National Research Fund (a member of Qatar Foundation). The findings achieved herein are solely the responsibility of the authors.

References

Bel-Air, F.: Demography, migration, and labor market in Qatar—Gulf Labor Markets and Migration (GLMM). European University Institute (EUI) and Gulf Research Center (GRC) (2014). Retrieved from http://cadmus.eui.eu/bitstream/handle/1814/32431/GLMM_ExpNote_082014.pdf?sequence=1

Colliers International.: Real Estate Market Overview Report—Doha, Qatar, Q2 2013. Colliers International (2013). Retrieved from http://www.colliers.com//media/files/emea/uae/market-overview/ci_qatar_re_overview_q2_2013.pdf

Colliers International.: Doha's Residential Market—Market Performance, Trends, and Affordability (2014). Retrieved from http://www.colliers.com//media/562F03BEE6D54CA0B7BA140DF287C911.ashx?la=en-GB

Dawkins, C., Koebel, C.: Overcoming barriers to placing manufactured housing in metropolitan communities. J. Am. Plann. Assoc. **76**(1), 73–88 (2009). https://doi.org/10.1080/01944360903401052

Eddisford, D., Carter, R.: The vernacular architecture of Doha, Qatar. J. Soc. Post-Medieval Archaeol. **51**(1), 81–107 (2017). https://doi.org/10.1080/00794236.2017.1320918

Haybatollahi, M., Czepkiewicz, M., Laatikainen, T., Kytta, M.: Neighborhood preferences, active travel behavior, and built environment: an exploratory study. Transp. Res. Part F Traffic Psychol. Behav. **29**, 57–69 (2015). http://dx.doi.org/10.1016/j.trf.2015.01.001

Hunt, J.: Stated preference examination of factors influencing residential attraction. In Pagliara, F., Preston, J., Simmonds, D. (eds.) Residential Location Choice: Advances in Spatial Science, pp. 21–59. Berlin: Springer (2010). ISBN:978-3-642-12788-5

Jaidah, I., Bourennane, M.: The History of Qatari Architecture 1800–1950. 1st edn. Skira (2009). ISBN-10:8861307930, ISBN-13:978-8861307933

Lee, H., Beamish, J., Carucci Goss, R.: Location preferences of multifamily housing residents. Hous. Soc. **35**(1), 41–58 (2008). https://doi.org/10.1080/08882746.2008.11430557

Ministry of Development Planning and Statistics.: Qatar Atlas, 3rd edn. Ministry of Development Planning and Statistics (2013). ISBN: 99921-62-95-3

Ministry of Development Planning and Statistics.: The Simplified Census of Population, Housing and Establishments, 2015. Ministry of Development Planning and Statistics (2015). Retrieved from https://www.mdps.gov.qa/en/knowledge/Publications/Other/Census%202015.pdf

Ministry of Municipality and Environment.: Archive Department. Doha Satellite Imagery in 2015 (2015)

Ouwehand, A., Doff, W.: 'What is the use of lifestyle research in housing?' A case study from the Netherlands. Paper presented at the 23rd conference of the European Network for Housing Research ENHR, France (2011). Retrieved from https://repository.tudelft.nl/islandora/object/uuid%3A1680568d-dd40-4792-a14a70b694b391cb. UUID 1680568d-dd40-4792-a14a-70b694b391cb

Petkar, A., Macwan, J.: Residential location preferences: approaches and research avenues. Paper presented at the 4th international conference on advances in civil engineering, AETACE (2013). Retrieved from http://searchdl.org/public/book_series/elsevierst/6/21_1.pdf

Pisman, A., Allaert, G., Lombaerde, P.: Urban and suburban lifestyles and residential preferences in a highly urbanized society. Belgeo **2**(1), 89–104 (2011). https://doi.org/10.4000/belgeo.6394

Property Market Overview.: Musheireb Properties (2017). Retrieved from http://www.msheireb.com/enus/aboutqatar/propertymarketoverview.aspx

Salama, A., Wiedmann, F.: Demystifying Doha: On Architecture and Urbanism in an Emerging City, 1st edn. Ashgate (2013). ISBN:978-1-4094-6634-5

Scheiner, J., Kasper, B.: Lifestyles, choice of housing location and daily mobility: the lifestyle approach in the context of spatial mobility and planning. Int. Soc. Sci. J. **55**(176), 319–332 (2003). https://doi.org/10.1111/1468-2451.5502011

Shandas, V., Makido, Y., Ferwati, S.: Rapid urban growth and land use patterns in Doha, Qatar: opportunities for sustainability? Eur. J. Sustain. Dev. Res. **1**(2), 1–13 (2017). https://doi.org/10.20897/ejosdr.201711

Snoj, J.: Population of Qatar by Nationality. Priya DSouza Consultancy (2017). Retrieved from http://priyadsouza.com/population-of-qatar-by-nationality-in-2017/

Wiedmann, F., Salama, A., Thierstein, A.: Urban evolution of the city of doha: an investigation into the impact of economic transformations on urban structures. METU J. Fac. Archi. (JFA) **29**(2), 35–61 (2012). https://doi.org/10.4305/metu.jfa.2012.2.2, Online at: http://jfa.arch.metu.edu.tr/archive/02585316/2012/cilt29/sayi_2/35-61.pdf

A Review on Prefab Industrialised Building System Modular Construction in Malaysia: The Perspective of Non-structural Studies

Teck Jung Chai[1(✉)], Cher Siang Tan[2], Tio Kiong Chow[3], Philip Chie Hui Ling[1], and Heng Boon Koh[4]

[1] Jabatan Pendidikan Politeknik & Kolej Komuniti, Aras 5, Galeria PjH, Jalan P4w, Persiaran Perdana Presint 4, 62100 Putrajaya, Wilayah Persekutuan, Malaysia
tjchai2@live.utm.my; tjchai7607@gmail.com
[2] Forensic Engineering Centre, Universiti Teknologi Malaysia, 81310 Skudai, Johor, Malaysia
[3] Politeknik Metro Betong, 1-12, Phase 2, Bandar Baru Jalan Baru, 95700 Betong, Sarawak, Malaysia
[4] Faculty of Civil and Environmental Engineering, Universiti Tun Hussein Onn Malaysia, 86400 Parit Raja, Johor, Malaysia

Abstract. Prefab modular construction approach is a well-established volumetric structural product through off-site prefabrication at factory. Modular construction approach builds confidents to promote sustainability in construction industry development and is beneficial to the environment ecosystem. The study explored and discussed on various non-structural perspective of IBS modular construction in Malaysia. The study aims to investigate the importance of the implementation of IBS modular construction in Malaysia as rapid efficient construction process and poised to help the Malaysia construction industry grow to solve the housing demand impacts. The study presented the definition, prefabrication technology, history, benefits, barriers and issues, application and characteristics of modular construction in the Malaysian construction industry through literature review. Based on the literature reviewed, the existing IBS practices in Malaysia is still low in application and modular construction practice currently had gained popularity worldwide and been proven as rapid and efficient construction method in IBS trend. Therefore, the existing IBS practices in Malaysia must be strategized towards more efficient and rapid construction method due to its numerous benefits and characteristics. The findings of this study will contribute to the transformation of existing construction practices to modern modular construction approach and proposed the future study on modular construction in structural perspectives.

Keywords: Prefab modular construction · Industrialised building system
Volumetric · Off-site prefabrication · Rapid

M. Awang and M. H. Isa (eds.), *The Advances in Civil Engineering Materials*,
Lecture Notes in Civil Engineering 19, https://doi.org/10.1007/978-981-13-2511-3_2

1 Introduction

Prefab is an architecture in building construction industry which had gained its popularity worldwide. Prefab is an evolutionary process but not revolutionary that was based on the successful and unsuccessful experiences [1]. The prefab history started in 1624 which first prefab house was completed and transported from England to Cape Ann [2, 3]. Modular construction is defined as off-site prefabrication in three-dimensional volumetric construction and modular building system where the structural components prefabricated by manufacturer and transported to site in ready-cut building components for erection [4–9]. Modular construction was used as the soldier's portable shelter solution during World War II and as intensive demand of housing after World War II in the United State since it was developed in 1940s [6, 8, 9]. Modular construction is based on rapid construction concept comprised of various technologies in two orders namely 2D frame system and volumetric elements in 3D [8].

IBS is Malaysia's local term to describe the use of automation, mechanisation and prefabrication of components for the local building industry [9]. IBS is characterized by its higher quality, faster jobs, increased profitability and the construction process which adopt prefab concept using standardised building components that are mass-produced in a factory ready for installation, transported and erected at site using suitable machinery and equipment with minimal workers with proper connection system [9–13]. The implementation of modular construction is proven to improve productivity, quality, efficiency, safety and health, speed up project duration, flexible, economically and promotes sustainable construction industry [5]. Prefabrication, pre-assembly, modularization and IBS are terms used to describe technologies advancement for rapid building construction [10].

This study aims to outline the importance of the implementation of IBS modular construction in Malaysia through the literature review. As the researcher's concern, the current implementation of IBS status in Malaysia construction project is low, despite its numerous benefits [9, 14]. The current IBS practices in Malaysia are more focused on piece-by-piece erection method for precast building components and then followed by services installation at site. Moreover, the revolution in building construction industry nowadays has transformed to the precast 2D or 3D volumetric rapid modern construction method which consists of various technologies. Due to this, the industry's mindset towards IBS modular construction in Malaysia must be changed [14]. The Malaysia government had driven the strategic planning in providing training and skills to the stakeholders. It is hoped that this will help to improve the workers knowledge, productivity and skill in modular construction adoption and thus benefits the stakeholders eventually. The rapid construction ignores the manufacturer time while site erection time is counted into construction period. The erection period of precast modules in complete volumetric unit is highly subjected to the joint design and erection principle. It is hoped that the current construction industry practices can be strategized and improved towards the government expectation with rapid, efficient and sustainable delivery construction environment.

This paper is expected to benefit the Malaysia construction industry in facing the housing demands issue within a short period at urban areas. This will help in improving the existing practice at construction site. The study will also contribute the new idea in precast IBS modular construction as sustainable delivery construction approach since this construction method is getting popular and capable to speed up the construction project completion period. Recent researchers' had come out with supporting fact that modular construction is an efficient construction process and poised to help the growth of construction industry [15]. This will help in transforming the existing IBS practice to the more efficient and rapid modular construction practice which will benefit the community and stakeholders.

2 Prefabrication Technology in Malaysia Construction Industry

IBS is the construction technologies, either offsite or onsite, involving the process of planning, designing, fabricating, transporting and assembling of building components, elements or modules for rapid assemble faster than traditional on-site construction with reasonable cost and quality [16–19]. This is an integration approach within the construction process which includes the production of prefabricated components, transportation of components and site erection techniques [20]. The IBS characteristics refer to open system, closed system, modular coordination, standardisation and tolerances, mass production, specialisation, stakeholders integrated coordination, production facility development, transportation, machineries and site erection equipment [10, 21]. The final products include componentized, panelised, and modularized elements [17]. Constructability, site logistics and fabrication process are the design concerned for off-site construction [19]. Offsite construction is that the prefabrication elements or parts of structures constructed in factory, then transported and assembled on-site while onsite construction is that the building blocks and parts of structures takes place directly on site [16]. There are two main IBS classes namely open system which standardized components, flexibility and design diversity, and closed system, which the element need specific made in factory [13]. In structural aspect, it can be divided into five key aspects namely precast concrete framing, panel and box systems, formwork systems, steel framing systems, prefabricated timber framing systems and block work systems [13]. The IBS classes, identification, structural aspects, systems definition and utilisation of IBS is summarised in Table 1.

The IBS implementation since early 1960s unfortunately found to be inappropriate for Malaysia's climate due to imported materials and method used will cause leakage and gaps in walls that came to light and currently the construction using IBS is slower than expected [9, 14, 22, 23]. Thus, in 1990s, prefabricated components had been applied to IBS as new approach and recently followed by sustainable green materials after Paris climate summit 2016 as specified in second IBS roadmap for sustainability, quality, efficiency and competency achievement [9, 24]. The Industrialised House Technologies was launched to lay down the scene of Modular IBS in Malaysia [5]. Due to the numerous benefits, Malaysian IBS suppliers and construction companies have created their own project teams for this industry, investment and innovation work [9]. Recent developments in modular volumetric or 3D printing have transformed to precise customisation of components as ideal construction industry with cheaper price and big size volume [9]. In September 2015, the Construction Industry Transformation Programme (Year 2016–2020) was launched in Malaysia with four strategic thrusts in which productivity was highlighted as significance key component of IBS [9, 25].

Table 1. Industrialised Building System (IBS) classes, identification, structural aspect/feature, system definition and IBS utilisation components [2, 6, 9, 11, 13, 16, 17, 21, 22, 23]

IBS classes	IBS identification	IBS Group (structural classification)	IBS modular construction	IBS system definition	IBS utilisation components
Off-site (open system)	Modular construction	Precast concrete framing, panel and box systems	Permanent	Volumetric, standardised 3D/room size units. Mass produced in factory, less waste and high-quality modules. Multiple sectional modules (6 sided boxes), transported and erect at site by crane	Volumetric, pod construction, 2D frames and 3D components, permanent concrete formwork
	Hybrid construction				Partial/semi-volumetric construction
	Panelised construction			Incorporated construction techniques with advance technology in factory, quality materials, less time, transported and erect on site	Open and closed panels, structural and composite non-structural insulated panels, prefabricated lightweight and roof panels
	Prefab timber framing systems from natural resources	Prefabricated timber framing systems		Offsite prefab components, using machinery, moulds and other mechanical equipment and erect at site	Timber frame and multilayer engineered timber product and recycled materials components
	Lightweight facades	Block work system		Casting concrete in a reusable mould, cured in a controlled environment, transported and lifted into place at site. Prestressed concrete is a method of using prestressing tendons that balance the tensile stress that concrete compression member would otherwise experience due to a bending load	Masonry blocks wall with timber and metal frame and ventilated facade system
	Sub-assemblies and accessories systems	Steel framing system			Floor or roof cassettes, precast concrete foundation assemblies, pre-assembled products and interlocking concrete masonry units, lightweight concrete blocks
On-site (closed system)	Prefabricated secondary structures assembled at site	Steel formwork system, steel framing system, block work system	Relocate-able	Offsite fabrication practice of building components in a factory or others manufacturing site to a greater degree of finish as bulk building structures and systems, and transporting complete assemblies or sub-assemblies to the construction site where the structure is to be located	Tunnel form, beams and columns moulding forms, permanent steel formworks (metal decks), stick build timber frame, insulated concrete formwork, thin joint blockwork/clay block

2.1 Characteristics of Modular Construction

Modular construction currently is the most capable construction technique, by combining various technologies to achieve, rapid construction and automation in construction industry development worldwide [10]. Modular construction is an off-site fabricated modules process in three dimensional volumetric units, modular building or system under controlled environment, same materials and design codes as conventional built facilities in modules form for site built facility when integrated on site [2, 6, 10, 15, 18, 26]. Modules involve three dimensional volumetric or box sections, multi section units, and stack-on units by using light gauge steel frame, timber frame, concrete, and composites materials that are commonly referred to pre-assembled pods, and been designed to be easily disassembled, moved and repositioned, where the module size is a factor of module location in the building, manufacturing constraints, and transportation limitations [2, 18].

Modularity concept consists of three different perspectives that are product, process, and supply chain between construction and manufacturing [15]. In construction industry, product refers to mix of components and spatial voids, module not complex and the supply chain is temporary multi-organizations for specific purpose of delivering a construction project [15]. Modular construction is widely used in low-rise buildings and recently multi-storey and even in high-rise construction with reinforced concrete or metal frame [10]. Modular high-rise buildings require a central structural core with three types of structural solutions, which are fabricated modular structure similar to ordinary modulus, fabricated reinforced concrete and composite system [10]. Modularity products is the key elements of mass customisation strategies as it uses a set of similar components modules to build a wide range of dwellings to improve the house building industry performance [15].

The application of modular construction can be either in permanent or relocate-able construction. Moreover, it is an innovative, offsite, lean manufacturing techniques, higher quality control, single or multi-story modules section, integrated or stand alone system as turn-key solution with the delivered forms in mechanical, electrical and plumbing fixtures and interior finishes [15]. This construction method produces less waste, faster completion time and sustainable construction delivery [6, 15]. Modular buildings are relocate-able, moveable and flexible building which is a partially or completely assembled building complies with applicable codes or state regulations that constructed in a building manufacturing facility by modular construction process [6, 15]. It is designed to be reused or repurposed multiple times and transported to different building sites and utilized for temporary space need [6, 15]. Thus, the raw materials demand and energy created to meet new need was reduced.

There is a need for early consideration for logistic transportation, storage and erection planning due to the size and weight of modular unit. The modular unit is not practical for storage and but must be installed immediately to the designated location [6]. Modular construction eases renovation work by just selecting and adding suitable modular units for the renovation project, and then installing the selected units into the existing building [6]. Besides, the well coordination, planning, and communication at the project early stage between the stakeholders must be developed to obtain good result for the final project. This is because any amendment in the middle stage of the project including services, mechanical and electrical parts will cause problem to the modular construction and will be costly.

Beside, the connections system between the modular units must be carefully considered by the parties that are involved the project at the early stage in design [6]. This is because the overall stability of the modular unit is subjected to the joint design. Due to this, the joint rigidity and their performance in structural perspective must be inspected both in the factory and on-site for safety. As engineer concerned, the ease of joint method used will influence the construction completion time where the erection time at site is encountered for rapid modular construction nowadays in construction industry worldwide. The modular construction significantly reduces site disruption, and vehicular traffic, and improves overall safety and security due to approximately 80% of the building construction activity done in factory [2, 15]. Construction of modular buildings occurs simultaneously with building site work, allowing projects to be completed in half of the time required by traditional construction method [6]. The comparison between modular construction and conventional construction schedule was shown in Fig. 1 [6, 15].

Structurally, the modular buildings are stronger than site-built construction because each module is engineered to independently withstand the rigors of transportation and craning onto foundations [15]. Manufacturing facilities have strict quality control programs with independent inspection and testing protocols that promote superior quality of construction [6]. The modules become one integrated wall, floor, and roof assembly once been put together and sealed [15]. Building offsite method ensures

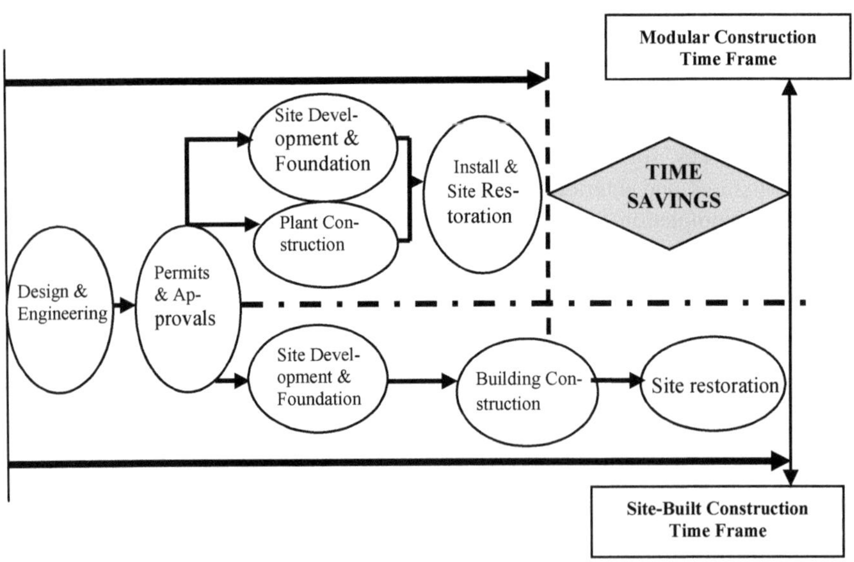

Fig. 1. Comparison between modular construction and conventional construction schedule [6, 15]

better construction quality management. There are three types of modular construction methods applied in high rise building construction namely core method, combination of core and podium method and modular in-fill method [27].

2.2 Prefab IBS Modular Construction Benefits, Issues and Challenges in Malaysia

Nowadays, modular construction naturally achieves the sustainability in the built environment with numerous applications of modular building solutions which benefits from the ease of use and flexibility [16, 28]. Beyond quality management and improved completion time, modular construction offers numerous other benefits to owners which include removing approximately 80% of the building construction activity from the site location [22]. The benefits of modular construction as mentioned by previous researchers are speed assembled [2, 6, 10, 17, 22, 29], enhanced productivity [6, 16, 29], high quality and safe working environment [2, 10, 16, 17, 22, 29, 30], better architectural appearance [16], vehicular traffic [22], new technologies testing [10, 30], innovative [22], efficient and cost-effective [16, 17, 22], rapid construction [10, 16], reduce overall cost [2, 16, 29], greater flexibility and reuse [22], enhanced durability [16], better engineered building and Building Information Modelling [22], economies [27], customization [27], environmental friendly and sustainability [2, 6, 10, 17, 22, 29, 30].

The adoption of prefabricated components has faced barriers such as the unattractive perception from the public, unstable and difficulty of coordinating delivery system, lack of integration into the design process and the use of shipping containers in modular construction has led to the misconception of modular construction [9]. The prefabricated modular structural components will be influenced by control distortion and buckling, transportation and manufacturing [19]. The lack of consensus on the conceptualization of product modularity and the differences between the construction and manufacturing industries were two main difficulties identified in the dissemination of modularity concept in the building construction sector [15]. This issue indicated that the importance of coordination, communication and planning in the early stage between the stakeholders to avoid amendment in the middle stage and reduce the project cost and hence convince the stakeholders adopt modular construction.

The implementation of prefabrication technology in Malaysia housing sector must be enhanced and strategized due to the low level of application in construction industry which was 42% of government projects and 70% of private projects in 2015, 69% government projects and 14% private sector [23–26, 31]. High initial cost, low profit margin, leakages, low quality, production delay, logistic problem, coordination, unskilled workforce and components erection problem were the barriers identified for Industrialised Building System modular construction implementation [31]. Modularisation and standardisation in building processes is currently keen promoted by Malaysia Construction Industry Development Board [9]. The education, training, computerisation and robotics currently play important roles in construction industry. According to A. Idris, the Industrialised Building System adoption issues and barriers revolve around the lack of expertise, transitional shift, financial, limited enforcement and industrialization objective [23]. Due to this, tax incentive, introduce quality control, research and development, and training programmes to increase uptake were introduced [9].

The challenges faced by the communities worldwide are more on the quality housing. According to Elena, there are 26 out of 34 megacities worldwide facing urbanisation wave, slums and unplanned development [10]. The Malaysian government has targeted 4,964,560 units of houses to be built between the years 1995 and 2020 [26]. In year 2005, the statistics showed that every third habitant of a city lived in unfavourable conditions and projected about 35 million apartments a year to be built to fulfil the urban housing demand [10]. In Malaysia, the population growth and urbanisation will lead to high demand of affordable housing issues [32]. Based on these facts, there is a need to change the existing IBS practices to the proven efficient and rapid modular construction as proved by China to construct fifty-seven story skyscrapers apartment within 19 days in 2015 [2].

The urban population growth in Malaysia is projected approximately to be 74.3% of total population in 2015 which will lead to an increase in demand for housing, infrastructure, education, and health facilities at urban area [33]. Moreover, the Department of Statistics Malaysia in 2016 has projected that the Malaysia population growth will achieve 41.5 million in year 2040 with 37 million and 87% Malaysian live in urban areas by 2050 [34, 35]. This statistics projection figures show the impact on housing demand and rapid construction method is needed to overcome it. The demographic research analysis carried out by the Malaysia Central Bank annual report in 2015 mentioned on the unaffordable housing price increase in all major cities in Malaysia [36, 37]. Therefore, prefab modular construction will serve as the new direction of construction technology to produce houses within short period.

3 Discussions

Based on the reviewed findings, previous Malaysia researcher's was more focused on the non-structural perceptive of IBS construction. The adoption of IBS approach in Malaysia is still low even though it has numerous benefits to the construction industry. Because of this, the implementation need to be enhanced and strategized to be subjected to local climate and issues identified. The Malaysia population growth facts indicate that the high demand of housing trend at urban areas will be the challenges for local construction industry. The existing construction practices cannot afford the high housing demand based on current construction ability trend. Due to this, the existing IBS practices need to be transformed toward a more efficient and rapid construction approach by using modular construction and Malaysia government had launched the Industrialised House Technologies to lay down the scene of IBS modular construction in Malaysia. China had adopted the modular construction and proved to complete the fifty-seven storeys, 800 unit skyscraper apartments within 19 days [2]. This will be the efficient and rapid construction model in precast construction industry worldwide.

The prefabrication technology with modularity and standardisation in building processes will be superior in providing the safe and quality houses within the short period and this method is flexible, sustainable delivery environment and relocate-able. Modular construction is a proven IBS new technology with the prefab modules done approximately 80% in factory with services attached and erect at site to the specific location with crane. The rapid completion of the project is encountered by the erection time at site while the prefab time in factory is normally ignored. The prefab modular construction must have efficient connection system to allow ease erection between the modules. Therefore, the joint of the precast modules must be carefully designed to produce safe and withstand modular unit totally. Due to this concern, there is a need to conduct the future study on the modular construction joint in structural perspective.

4 Conclusion

Based on the literature review findings, these are the conclusions drawn for the studies as stated below;

1. Prefab IBS modular construction technique was proven efficient, rapid and sustainable construction delivery.
2. The existing IBS practice adoption in Malaysia has to be im proved, transformed and strategized to achieve better construction project outcome.
3. Communication, planning and coordination between the stake holders in project early stage is the key success factor for project success.
4. Structural perspective study needed to provide safe, withstand and more comprehensive modular construction system in future.

Acknowledgements. The authors would like to express their appreciation for the support of Malaysia Ministry of Higher Education (FRGS-4F834) and Universiti Teknologi Malaysia (GUP 13H37 and 03G55).

References

1. Smith, R.E.: Prefab Architecture: A Guide to Modular Design and Construction. Wiley, New York (2011)
2. Fred, E.B., Kim, J.H., Kim, J.T.: Performance of modular prefabricated architecture: case study-based review and future pathways. Sustainability **8**(558) (2016)
3. Tomas Jr., U.G., Almarwae, M.: Prefabricated technology in a modular house. Int. J. Adv. Sci. Technol. **73**(2014), 51–74 (2014)
4. Musa, M.F., Mohammad, M.F., Yusof, M.R., Ahmad, R.: Industrialised Building System Modular System (IBSMS) organisational readiness framework. Procedia Soc. Behav. Sci. **222**(2016), 83–92 (2016)
5. Musa, M.F., Mohammad, M.F., Mahbub, R., Yusof, M.R.: Enhancing the quality of life by adopting sustainable modular Industrialised Building System (IBS) in the Malaysian Construction Industry. Procedia Soc. Behav. Sci. **153**(2014), 79–89 (2014)
6. Musa, M.F., Yusof, M.R., Mohammad, M.F., Samsudin, N.S.: Towards the adoption of modular construction and prefabrication in the construction environment: a case study in Malaysia. ARPN J. Eng. Appl. Sci. **11**(13), 8122–8131 (2016)
7. Lawson, R.M., Ogden, R.G., Bergin, R.: Application of modular construction in high-rise buildings. J. Archit. Eng. **18**(2), 148–154 (2012)

8. Rostami, R., Lamit, H., Khoshnava, S.M., Rostami, R.: Industrialization and sustainable construction a case study of Malaysia. Asian J. Microbial. Biotech. Environ. Sci. **15**(2), 433–440 (2013)
9. Construction Industry Development Board (CIDB).: Industrialised building system: the path to enhanced productivity. Heights **3** (2016). Retrieved from website http://www.cidb.gov.my/images/content/pdf/Korporat/Heights3-compressed.pdf on 12 Dec 2017
10. Elena, M.G., Viktor, P.G., Anna, A.K.: Modular buildings in modern construction. Procedia Eng. **153**(2016), 167–172 (2016)
11. Musa, M.F., Mohammad, M.F., Yusof, M.R., Mahbub, R.: The Way Forward for Industrialised Building System (IBS) in Malaysia. In: Proceedings of the International Civil and Infrastructure Engineering Conference 2014 (InCIEC 2014), vol. 14, pp. 163–175, Springer, Singapore (2015)
12. Mohammad, M.F.: Construction environment: adopting IBS construction approach towards achieving sustainable development. Procedia Soc. Behav. Sci. **85**(2013), 8–15 (2013)
13. Nawi, M.N.M., Azman, M.N.A., Baluch, N., Kamar, K.A.M., Hamid, Z.A.: Study on the use of industrialised building system in malaysian private construction projects. ARPN J. Eng. Appl. Sci. **10**(17), 7368–7374 (2015)
14. Lee, R.: IBS Adoption in Malaysia Remains Low (2015). Retrieved from website http://www.theedgeproperty.com.my/content/ibs-adoption-malaysia-remains-low-cidb on 16 Dec 2017
15. Modular Building Institute (MBI).: Why Modular? (2017). Retrieved from website http://www.modular.org/HtmlPage.aspx?name=why_modular on 22 Dec 2017
16. Lenka, K., Tomas, M., Peter, M.: Modern methods of constructions and their components. J. Eng. Archi. **2**(1), 27–35 (2014)
17. Ryan, E.S.: Off-Site Construction Implementation Resource: Off-Site and Modular Construction Explained. National Institute of Building Sciences (2014). Retrieved from website https://c.ymcdn.com/sites/www.nibs.org/resource/resmgr/OSCC/OSMC_Explained.pdf on 18 Nov 2017
18. Seffar, N., Abbott, C.: The Use of Off-site Construction (OSC) Method to Reconstruct Iraq. In: Proceedings of the 13th International Postgraduate Research Conference 2017, University of Salford, Manchester, 14–15 Sept 2017 (2017). Retrieved from website http://usir.salford.ac.uk/43913/1/13th%20IPGRC%202017%20Full%20Conference%20Proceedings.pdf on 10 Dec 2017
19. Salama, T., Salah, A., Moselhi, O.: Configuration of hybrid modular construction for residential buildings. Int. J. Innov. Manag. Technol. **8**(2), 106–112 (2017)
20. Fathi, M.S., Abedi, M., Mirasa, A.K.: Construction Industry Experience of Industralised Building System in Malaysia. In: Proceedings of the 9th International Congress on Civil Engineering, May 8–10, 2012 Isfahan University of Technology (IUT), Isfahan, Iran (2012). Retrieved from website https://www.researchgate.net/publication/307932416_Construction_Industry_Experience_of_Industralised_Building_System_in_Malaysia on 10 Dec 2017
21. Richard, R.B..: Industrialised building systems: reproduction before automation and robotics. Autom. Constr. **14**(4), 442–451 (2005)
22. Amin, M.A.M., Abas, N.H., Shahidan, S., Rahmat, M.H., Suhaini, N.A., Nagapan, S., Rahim, R.A.: A review on the current issues and barriers of Industrialised Building System (IBS) adoption in Malaysia's construction industry. Mater. Sci. Eng. **271** (2017). https://doi.org/10.1088/1757-899x/271/1/012031
23. Idris, A.: Minister: IBS to be compulsory for construction projects in 2018. Edge Markets (2016). Retrieved from website http://www.theedgeproperty.com.my/content/1099599/minister-ibs-be-compulsory-construction-projects-2018 on 15 Dec 2017

24. IBS Roadmap 2011–2015.: Construction Industry Development Board (CIDB). Kuala Lumpur, Malaysia (2010)
25. Mydin, O., Sani, N.M., Taib, M.: Industrialised Building System in Malaysia: A Review. MATEC Web Conferences, vol. 10, issue no. 1002 (2014). http://doi.org/10.1051/matecconf/20141001002
26. Hyung, K.P., Jong, H.O.: Unit modular in-fill construction method for high rise building. KSCE J. Civ. Eng. **20**(4), 1201–1210 (2016)
27. Rocha, C., Formoso, C., Tzortzopoulos, P.: Adopting product modularity in house building to support mass customisation. Sustainability **2015**(7), 4919–4937 (2015)
28. Yong, H.A., Kyoon, T.K.: Sustainability in modular design and construction: a case study of 'The Stack'. Int. J. Sustain. Build. Technol. Urban Dev. **5**(4), 250–259 (2014)
29. Jalil, A.A., Nuruddin, A.R., Jaafar, M., Mydin, M.A O.: A New Procurement Method for Housing Projects Implementing IBS Modular System. In: Proceedings of International Conference on Advances in Civil and Environmental Engineering 2015, pp. 1–8 (2015)
30. Zhang, X.X., Wang, L., Jiang, P.M.: Comprehensive Benefits Analysis of Steel Structure Modular Residence Based on the Entropy Evaluation. In: Proceedings of The 3rd International Conference on Energy Materials and Environment Engineering, Earth and Environmental Science, vol. 61 (2017)
31. Aris, N.A.M., Fathi, M.S., Harun, A.N., Mohamed, Z.: Prefabrication Technology for Sustainable Affordable Housing Supplky in Malaysia. In: E-Proceedings of1st International Conference on Sustainable Infrastructure and Engineering 2017 In Conjunction with UTMRS 1st International Conference On Advanced Technology 2017 (UTMRS-ICAT 2017), 10th Oct 2017, Renaissance Hotel, Kuala Lumpur (2017)
32. Shuid, S.: The housing provision system in Malaysia. Habitat Int. **54**, 210–223 (2015)
33. Economic Planning Unit.: Eleventh Malaysia Plan 2016–2020: Anchoring Growth on People. The Economic Planning Unit, Prime Minister's Department, Malaysia (2015). Retrieved from website http://www.epu.gov.my on 26 Dec 2017. ISBN: 978-9675842085
34. PR1MA.: Membina Komuniti Bersama Laporan Tahunan 2015. Kuala Lumpur (2015)
35. Department of Statistics Malaysia.: Population Projection (Revised), Malaysia 2010–2040 (2016). Retrieved from website https://www.statistics.gov.my on 29 Dec 2017
36. Saikah, M., Kasim, N., Zainal, R., Sarpin, N., Rahim, M.H.I.A.: Potential implementation of light steel housing system for affordable housing project in Malaysia. Mater. Sci. Eng. **271** (2017). https://doi.org/10.1088/1757-899x/271/1/012106
37. Annual Report Bank Negara Malaysia.: Financial stability and payment systems report 2014. Kuala Lumpur: Bank Negara Malaysia (2015). Retrieved from website http://www.bnm.gov.my/files/publication/fsps/en/2014/fs2014_book.pdf on 26 Dec 2017

A Review on Repair of Damaged Concrete Column Under Cyclic Loading by Using Confinement

Jia-Yang Tan[1(✉)], Chau-Khun Ma[1], Nazirah Mohd Apandi[1], Abdullah Zawawi Awang[1], and Wahid Omar[2]

[1] Department of Structure and Materials, Faculty of Civil Engineering, Universiti Teknologi Malaysia, Johor Bahru, Malaysia
jytan6@graduate.utm.my
[2] Office of Chancellery, Universiti Teknologi Malaysia, Johor Bahru, Malaysia

Abstract. The earthquake stroke Sabah in 2015 have given an alarm to structural engineers on the importance of ensure concrete structures to resist seismic load. The damaged column due to the seismic load needed to be repaired. Research and development of rapid repair techniques have emerged as one of the most critical researches globally as the money spent on repairing damaged structures are more than the cost of building new structures. Confinement is one of the popular and effective rapid repairing of damaged concrete under cyclic loading. This paper reviewed the uses of confinement as a repairing technique to facilitate the research development in this field. The findings have proven that confinement repair techniques are effective and can restore the original capacities of damaged concrete members.

1 Introduction

Nowadays, sustainable development has attracted public attentions in conjunction to save the environment [1]. Sustainable development is the development which fulfils the needs of the present generation without compromising the future generations to meet their needs. This is because the increasing of world population makes severe and excessive demands on the limiting available resources on the earth. Sustainable development is the effort to reduce the environment impact while being still able to achieve social and economic growth simultaneously.

Earthquakes all around the world have shown the vulnerability of existing reinforced concrete structures to moderate and strong ground motions [2]. The earthquake stroke Sabah on 2015 with Richter scale of 6.0 have alarmed the public about the importance of all the buildings and structures to resist seismic load. Bridge piers and building columns may experience complex axial, shear, bending, and torsional loads during an earthquake. The aftermath of the previous earthquakes was terrible and most of the building columns and bridge piers were damaged to different degree of damage condition as shown in Fig. 1.

M. Awang and M. H. Isa (eds.), *The Advances in Civil Engineering Materials*, Lecture Notes in Civil Engineering 19, https://doi.org/10.1007/978-981-13-2511-3_3

Fig. 1. The damaged column under seismic loading [4]

The damaged columns or bridge piers need to be assessed to ensure the serviceability of the structures after earthquake events. The columns or bridge piers were mostly going to be demolished due to the severity of damage conditions after an earthquake. However, it is not a sustainable way and it incurs huge cost to demolish every damaged column and bridge pier. In accordance with the sustainability development goals, some alternatives are needed to repair the structures so that the structures are safe to be used.

2 Literature Review

In recent decades, more researches aimed to repair damaged column under cyclic loading. There are several available techniques of repairing damaged structures including concrete jacketing, steel jacketing, fibre reinforced polymer (FRP) and steel strapping which have proven to be effective to restore the strength and ductility of the columns or bridge piers and make the structure back to service after damaged by cyclic loading.

The researches on cyclic loading usually carried out on column size specimens with a footing which demonstrates as a column with a fixed support in the laboratory with set up as shown in Fig. 2. The specimens were tested under lateral cyclic loading scheme to resemble seismic load. A constant axial load was applied to assemble the dead load from the superstructure.

An overview of existing studies on repairing damaged concrete column under cyclic loading using confinement is shown in Table 1. As early as 1997, Ghobarah et al. [2] carried out the researches on 3 full scale rectangular columns by using steel jacketing to

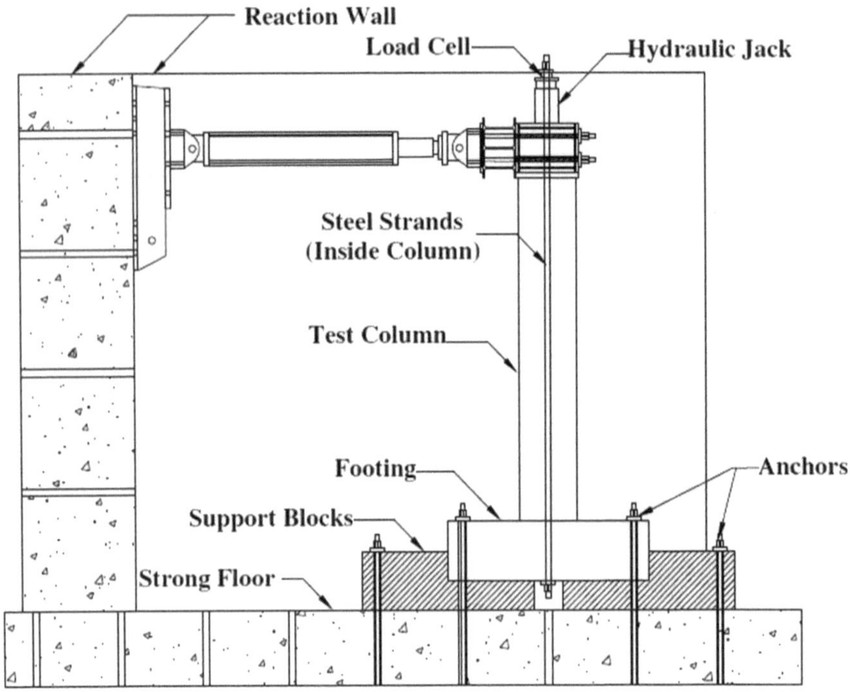

Fig. 2. The example of experimental set up [3]

repair the columns after damaged. They have gained positive results after rehabilitation. In 2000, K. Fukuyama et al. (2000) extended their studies by using two more different methods of repairing, namely RC jacket and carbon fibre sheets jacket to repair the columns after cyclic lateral load had been induced. They altered the damage degree and cross section size and successfully restore ductility even though the damaged concrete remains untreated [5]. Cheng et al. (2003) used FRP with dog bone shaped bar to repair one full-scale square and three full scale circular bridge piers after loaded under constant axial load and cyclic lateral load. The fractured bars were repaired, and the deformability of the columns were enhanced by FRP wraps [6]. In 2006, Nesheli and Meguro (2006) extended the studies by using pre-tensioned carbon or aramid FRP belts with different shear-span to depth ratios. The result shown that the repaired columns indicated better seismic performance [7]. Li and Sung [8] furthered their researches by repairing columns using epoxy and non-shrinkage mortar and rehabilitated by CFRP in 2003 which showed that repaired columns have improved hysteretic responses at high displacement ductility significantly. Sun et al. (2011) used CFRP to repair six circular piers to enhance shear strength and lateral displacement [9]. In 2013, He et al. [3] altered the torque-to-moment ratios of the columns with different loading combinations, they restored the strength of the columns without fractured bars but they were only partially successful for the column with fractured bars located near the base using CFRP. Li et al. (2017) investigated the effect of column heights and repairing processes, either with or without axial loads to the load-carrying capacity and ductility of the repaired columns. The results showed that the performances of the repaired columns are better than those of the original columns [10]. Truong et al. [11] researched on the retrofitting strategies either

Table 1 An overview of previous studies on repair of damaged concrete column under cyclic loading using confinement

References	Loading scheme	Type of confinement	Confining material	Number	Sample scale	Shape	Testing parameter	Findings
Ghobarah et al. [2]	Constant axial load + cyclic lateral load	Steel Jacketing	Corrugated steel	3	Full scale	Rectangular	Transverse reinforcement and Jacketing of the column	Capable to rehabilitate the original under designed column
Fukuyama et al. [5]	Constant axial compressive load + cyclic shear forces	RC jacketing, steel plates jacketing, carbon fibre sheets jacketing	Reinforced concrete (RC), steel plates, carbon fibre (CF) sheets	8	1/2	Square	Cross section size, damage degree	Shear strength and ductility of the repaired columns, in which the concrete remained crushed and the longitudinal bars remained buckled, can be restored over the pre-damaged level
Cheng et al. [6]	Constant axial load + cyclic lateral load	FRP	Dog-bone shape bars to replace the fractured longitudinal bars in plastic hinges and using FRP wraps	4	Full scale	1 Square and 3 circular	Column cross section shape, height of column	The fractured longitudinal bars can be completely repaired and the deformation capacities of the columns were enhanced by FRP wraps. The results also indicate the repaired column strengths are less than anticipated due to concrete deterioration and the buckling of the longitudinal bars
Nesheli and Meguro [7]	Cyclic lateral loading and constant axial compression	FRP	Pre-tensioned carbon or aramid FRP Belts	5	1/2	Square	Shear span-to-depth ratios, type of FRP belt	When the damaged columns were retrofitted by pre-tensioned FRP belts, the lateral strength could be restored and the drop in shear capacity could be prevented up to large drifts, indicating a better seismic performance
Li and Sung [8]	Constant axial load + cyclic load	FRP	Epoxy and non-shrinkage mortar and rehabilitated by CFRP	1	40%	Circular	Jacketed/unjacketed	Repaired and rehabilitated column develops significantly improved hysteretic responses at high displacement ductility

(*continued*)

Table 1 (*continued*)

References	Loading scheme	Type of confinement	Confining material	Number	Sample scale	Shape	Testing parameter	Findings
Zhiguo et al. [9]	Constant axial load + lateral cyclic load	FRP	CFRP	6	1/3	Circular	Height of the pier, longitudinal bar ratio, spiral bar spacing	Shear strength and lateral displacement of the repaired piers increased
He et al. [3]	Increasing cyclic lateral loading + constant axial load	FRP	CFRP	5	1/2	Square	Damage condition, loading combination (torque-to-moment ratio)	Successful in restoring the strength of the columns without fractured bars, but partially successful for the column with fractured bars near the base
Li et al. [10]	Amplitude-increasing lateral loads and constant axial load	Concrete jacketing	High-performance Fibre-Reinforced cementitious Composite	4	–	Square	Repair column heights and repairing processes (with and without axial loads)	Load-carrying capacity and ductility of the repaired columns higher than those of the original columns. With axial loads during repairing, the repaired columns displayed better cyclic performance
Truong et al. [11]	Cyclic loading combined with axial loading	Steel and concrete jacketing, CFRP	Non-shrinkage mortar and new concrete jacketing with amorphous metallic fibre reinforced concrete	20	1/2	Square	Retrofit strategies (partial retrofit in the plastic hinge zone, and full retrofit in the entire range of columns)	Retrofitted specimens presented ductile failure mode and enhancement in the dissipated energy and the damping ratio, but the effect differed for each retrofit method
Parghi and Alam [12]	Cyclic lateral loading and axial loading	CFRP	CFRP	12	–	Circular	Strength of concrete, yield strength and amount of longitudinal steel rebar, level of axial load, shear span-to -depth ratio, and CFRP confinement layer	It was observed that the amount of reinforcement, shear span-depth ratio, and level of the axial load could significantly affect the collapse fragility curve of the retrofitted bridge piers

partial retrofitting in the plastic hinge zone to increase the deformability, or full retrofitting in of columns. Parghi and Alam [12] used CFRP to repair the circular piers damaged under cyclic lateral loading and axial loading. It was observed that the amount of reinforcement, shear span-depth ratio, and level of the axial load could significantly affect the collapse fragility curve of the retrofitted bridge piers.

3 Implemented Approaches for the Development of Confinement Repairing Techniques

The literature reviews have clearly indicated that the confinement repairing techniques can be roughly categorized into 3 main groups, which are concrete jacketing, steel jacketing and FRP. The most famous method is FRP confinement as there are many research works focused on it. In these researches, the column or bridge piers were damaged first and then repaired by confinement to check the effectiveness of these materials as confining agents through their restorability on the capacity, deformability, ductility and the stiffness of the structure members. The taxonomy of the research in this area is presented in Fig. 3.

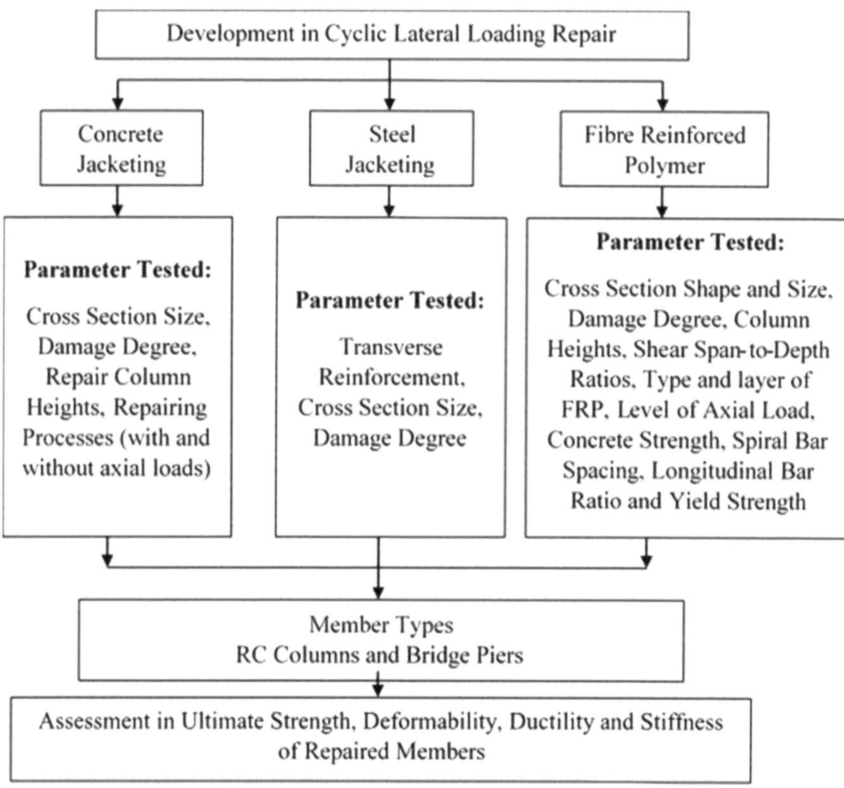

Fig. 3. Taxonomy of confinement repair technique researches on cyclic loading

4 Conclusion

This paper reviewed some of the methods and tests conducted to evaluate the efficiency of confinement repairing technique in repairing damaged concrete columns or bridge piers under cyclic loading. Several conclusions can be made from this review:

(a) The efficiencies of the proposed methods were evaluated by comparing the restored capacities of the damaged columns or piers. Normally the specimens were compared by assessment in ultimate strength, deformability, ductility and stiffness of repaired members.
(b) Past studies have covered a wide range of variables including cross section shape and size, damage degree, repair column height, repair process, transverse reinforcement ratio, shear-span-to-depth ratio, type and layer of FRP, spiral bar spacing, level of axial load, longitudinal bar ratio and yield strength. Most of the past studies have focused on testing the efficiency of several types of confining materials as repair techniques. In cyclic load tests, the pre-damaged condition level and slenderness ratio were quite critical.

Acknowledgements. The scholarship support from UTM Zamalah is highly appreciated.

References

1. Altwair, N. M., and Kabir, S.: Green Concrete Structure by Replacing Cement with Pozzolanic Materials to Reduce Greenhouse Gas Emission for Sustainability Environment. In: 6th International Engineering and Construction Conference (IECC'6). Cairo, Egypt (2010)
2. Ghobarah, A., Biddah, A., Mahgoub, M.: Rehabilitation of reinforced concrete column using corrugated steel jacketing. J. Earthq. Eng. **1**, 651–673 (1997)
3. He, R., Sneed, L.H., Belarbi, A.: Rapid repair of severely damaged RC columns with different damage conditions: an experimental study. Int. J. Concr. Struct. Mater. **7**, 35–50 (2013)
4. SabahKini, E.T.: Earthquake Damaged 31 Health Facilities In Sabah—Subramaniam (2015). Retrieved from SabahKini: ahkini.com/earthquake-damaged-31-health-facilities-in-sabah-subramaniam/
5. Fukuyama, K., Higashibata, Y., Miyauchi, Y.: Studies on repair and strengthening methods of damaged reinforced concrete columns. Cem. Concr. Compos. **22**, 81 ± 88 (2000)
6. Cheng, C.-T., Yang, J.-C., Yeh, Y.-K., Chen, S.-E.: Seismic performance of repaired hollow-bridge piers. Constr. Build. Mater. **17**, 339–351 (2003)
7. Nesheli, K.N., Meguro, K.: Seismic Retrofitting of Earthquake-Damaged Concrete Columns by Lateral Pre-tensioning of FRP Belts. In: Proceedings of the 8th U.S. National Conference on Earthquake Engineering, San Francisco, California, USA (2006)
8. Li, Y.F., Sung, Y.Y.: Seismic repair and rehabilitation of a shear-failure damaged circular bridge column using carbon fiber reinforced plastic jacketing. Can. J. Civ. Eng. (2011)
9. Sun, Z., Wang, D., Du, X., Si, B., : Rapid repair of severely earthquake-damaged bridge piers with flexural-shear failure mode. Earthq. Eng. Eng. Vib. **10**, 553–567 (2011)

10. Li, X., Wang, J., Bao, Y., Chen, G.: Cyclic behavior of damaged reinforced concrete columns repaired with high-performance fiber-reinforced cementitious composite. Eng. Struct. **136**, 26–36 (2017)
11. Truong, G.T., Kim, J.-C., Choi, K.-K.: Seismic performance of reinforced concrete columns retrofitted by various methods. Eng. Struct. **134**, 217–235 (2017)
12. Parghi, A., Alam, M.S.: Seismic collapse assessment of non-seismically designed circular RC bridge piers retrofitted with FRP composites. Compos. Struct. **160**, 901–916 (2017)

An Assessment on the Physical and Rheological Properties of Asphalt Binder Modified with Micro Bauxite Powder (MBP)

Hamza Mohamed Abdulhamid Elghatas[1(✉)], Mohamad Yusri Bin Aman[2], and Ahmed Suliman B. Ali[1]

[1] Faculty of Civil and Environmental Engineering, Universiti Tun Hussein Onn Malaysia, 86400 Parit Raja, Malaysia
alkatass@gmail.com

[2] Faculty Civil Environmental and Engineering, Smart Driving Research Centre, University Tun Hussein Onn Malaysia, Parit Raja, Malaysia
mdyusri@uthm.edu.my

Abstract. This investigation delved into the effect of micro bauxite powder (MBP) on the physical and rheological characteristics of asphalt binder. A high-shear mixer set at 4000 rpm was used to blend three varying percentages of MBP (3, 5 and 7% by weight of asphalt binder) with the asphalt binder. The base asphalt binder was left devoid of MBP. The physical properties of the asphalt binder merged with MBP were assessed through standard tests (to ascertain penetration, ductility and the softening point), and its rheological properties were gauged by means of a dynamic shear rheometer (DSR). While the standard tests revealed that the physical properties of the asphalt binder were improved by the MPB modifier, the DSR outcomes disclosed an upgrading in its (G*) and (δ) values at raised temperatures. These are indications that the viscoelastic features of the asphalt binders were enhanced by the modification. In comparison to the base asphalt binder, the modified binders exhibited an elevated degree of elasticity, stiffness and resistance against rutting deformation. Thus, the modification of asphalt binders using MBP can serve to improve the performance of pavements.

Keywords: Physical properties · Micro bauxite powder and dynamic shear rheometer

1 Introduction

Asphalt pavement is the most commonly used material around the world for the paving of roads. The performance and resilience of asphalt pavements are affected by several issues. Among them are service circumstances (including drastic temperature changes and recurring extreme axle loading) and the quality of raw materials used [1]. Road pavements are fabricated to withstand external loads (including mechanical loading deriving from heavy traffic) and thermal loading arising from thermal alterations [2]. To realize good performance, highway authorities constantly harp on the use of specific materials for the construction of pavements. Asphalt modifiers have been used to

M. Awang and M. H. Isa (eds.), *The Advances in Civil Engineering Materials*,
Lecture Notes in Civil Engineering 19, https://doi.org/10.1007/978-981-13-2511-3_4

enhance the performance of asphalt pavements since the 1950s. The focus of efforts in this field is mainly directed at raising the resistance level against pavement distresses [3]. The results from previous studies indicate that bitumen modifiers such as fillers, extended polymers, fibres, oxidants, and nanoclay can be used for improving the properties of asphalt binders. Over the years, the use of mineral fillers for asphalt binding has been on the rise. The inclusion of fillers has been observed to not only enhance the features of the binder, but also raise the level of resistance against irreversible deformation.

Bauxite, a natural material resembling clay, varies in colour from white to dark brown or red. This variation in colour is dependent on the make-up and extent of its constituents. Bauxite comprises mainly one or more aluminium hydroxide minerals and a variety of combinations involving silica (SiO_2), titanium (TiO_2), iron oxide (Fe_2O_3), aluminosilicates (clay etc.), and several other barely detectable impurities [4].

Berthier, a Frenchman, discovered bauxite in 1821 while scrutinizing specimens uncovered close to Les Baux in the south of France. However, it was many years later that the term 'bauxite' came into being [5].

Bauxite mining in Peninsular Malaysia has been going on since 1936. During World War II, several small bauxite deposits were exhausted from mining operations in some parts of Johor and near the coast of Malacca [6]. According to a survey [7], Malaysia ranked third among the world's bauxite producers for the year 2015 (Fig. 1). However, the detrimental environmental effects linked to bauxite extraction, led to a

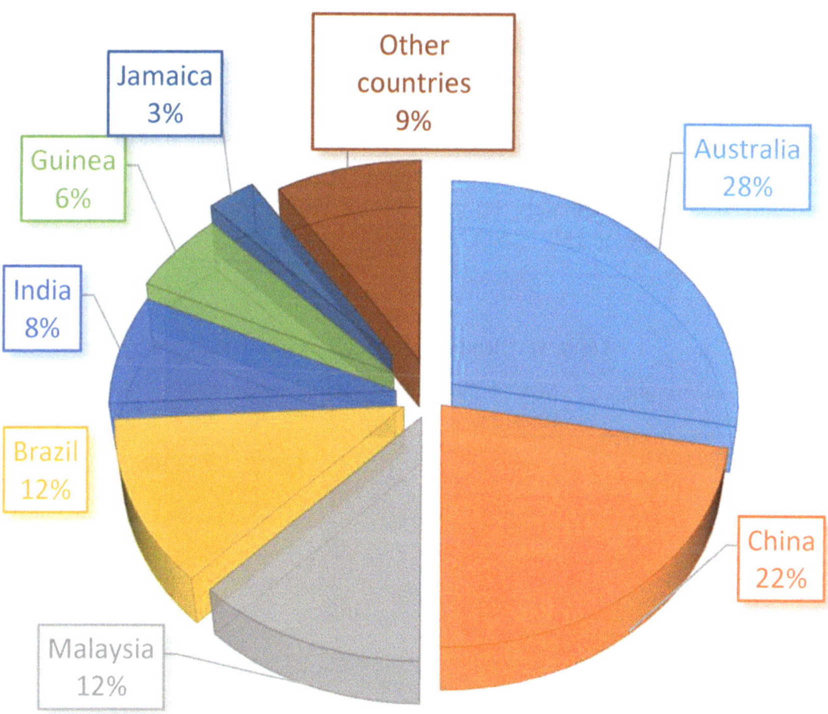

Fig. 1. Global bauxite production for 2015

cessation of its mining activities and export since the 15th of January, 2016. While the demand for aluminium is growing by leaps and bounds, the acknowledged reserves of its bauxite ore are deemed adequate for meeting global needs over the long term.

2 Experimental

This section explains activities on each of materials selection, laboratory equipment, standards and testing procedures that applied to accomplish the objectives of this study.

2.1 Materials

The control asphalt binder employed was bitumen with a penetration standing of 80/100. The binder, which was sourced from Petronas Company (Malaysia), met all the conditions set by Jabatan Kerja Raya (Malaysian standard). Malaysian bauxite, procured from KK Best Metal Sdn. Bhd. Kuantan, Pahang, Malaysia, was used to modify the bitumen binder.

2.2 Sample Preparation

The bauxite was first dried at a temperature of 155 °C to get rid of the moisture. Then it was ground into a specific powder size and sieved using a 75 μm sieve. Three levels of bauxite powder content (3, 5 and 7% by weight of bitumen) were used. The binder base was left devoid of bauxite powder.

A stainless steel container (600 ml) was filled with 400 g of 80/100 grade base bitumen and placed in a thermoelectric heater. When the temperature of bitumen reached 150 °C, a high-shear mixer was dipped into the container and set to 4000 rpm. 3, 5 and 7% of Malaysian bauxite (by weight of the asphalt) was added into the container with 0% MBP representing the control. This is illustrated in Table 1. Mixing was continued for approximately 60 min, during which the temperature was maintained within the range of 150 ± 5 °C.

Table 1. Blending procedure parameters

Binder	Total weight (gm)	Mixing speed (rpm)	Mixing temperature (°C)	Mixing time (min)
Control	400	4000	150	60
MBP 3%	412.4	4000	150	60
MBP 5%	421	4000	150	60
MBP 7%	430	4000	150	60

2.3 Physical Properties

The physical properties of a MBP modified asphalt binder is a significant factor during efforts to determine its performance level [8]. The characterization of the altered and unaltered asphalt binders was achieved through the conducting of standard tests to ascertain penetration, the softening point and ductility. All the specifications adhered to the ASTM standards. This is portrayed Table 2.

Table 2. ASTM test specifications

Physical properties	Specification	Limits
Penetration @ 25 °C (mm)	ASTM D5	80–100
Softening point (°C)	ASTM D36	45–52
Ductility @ 25 °C (cm)	ASTM D113	≥ 100

2.4 Rheological Properties

The dynamic shear rheometer (DSR) test was carried out for the purpose of establishing the rheological characteristics of the asphalt binders. This test was run from temperatures ranging from transitional to elevated. In order to determine the rheological properties of asphalt binders, the Dynamic Shear Rheometer (DSR) test was performed from intermediate to high temperatures. The DSR test serves to provide an approximation on the capacity of the asphalt binder to defy irreversible deformation. DSR test can estimate the resistance of the asphalt binder to permanent deformation. Initially, the complex modulus ‘G*’ and phase angle ‘δ’ were ascertained. This facilitated an examination of the asphalt binder’s resistance against irreversible deformation, and the evaluation of its performance under transitional and elevated temperatures. The test was conducted in accordance with AASHTO T315.

3 Results and Discussions

3.1 Physical Properties

The tests’ results for each of penetration, ductility and the softening point are discussed in this section, to investigates the effect of the modifier on the physical characteristics of asphalt binder.

3.1.1 Penetration

Penetration values are used to classify the material into grades. The higher the value, the softer the binder, but a workable binder is recommended [9]. It was observed that the penetration value of the modified asphalt binders with MBP was slightly lower than the base asphalt. This indicates that the increase in binder stiffness was caused by the addition of different percentages of MBP to the base asphalt binder. The average penetration value for the base asphalt binder was 94 mm. This places it in the 80–100 grade. The modified asphalt binder with MBP 3% was 89 mm, followed by MBP 5% with 87 mm and MBP 7% with 83 mm. This is displayed in Fig. 2.

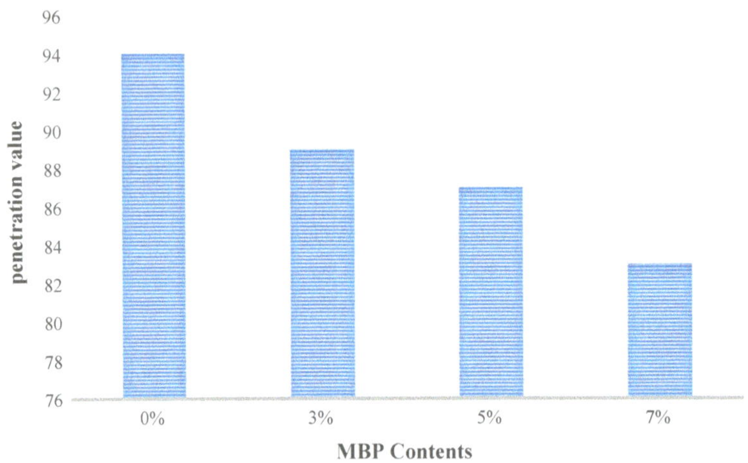

Fig. 2. Penetration of modified asphalt binders

3.1.2 Ductility

Due to its viscoelastic nature, the asphalt binder does not have a well-defined breaking point. Instead, it gradually transforms into a soft form as the temperature rises [10]. The purpose of the ductility test is to assess the flexibility of the asphalt binder at a temperature of 25 °C. This paves the way for a prediction on the asphalt binder's degree of flexibility when exposed to raised and reduced temperatures. The ductility test was carried out for the base and altered asphalt binders in compliance with ASTM D113. While the ductility value for the base asphalt binder was observed to be 140 cm, the values for the altered asphalt binders were 124 for MBP 3% 127 cm for MBP 5% and 101 cm for MBP 7% as shown in Fig. 3. It was perceived that the asphalt binder's ductility value decreases as the percentage of MBP increases. This is attributed to the progression in the rigidity of the asphalt binder.

3.1.3 Softening Point

The vulnerability of the asphalt binder to flow at an elevated temperature can be gauged by way of the softening point test. Put plainly, this test establishes the temperature at which the asphalt binder is altered from a solid to liquid state. The softening point test was performed for the base and the modified asphalt binder according to ASTM D36. The results revealed that the softening point of modified asphalt binders rose in tandem with the increasing addition of MBP, while the base bitumen displayed the lowest value in comparison to the blends. The softening point value for the base asphalt binder was 46 °C, while the modified asphalt binder with MBP 3% was 47 °C, MBP 5% was 48 °

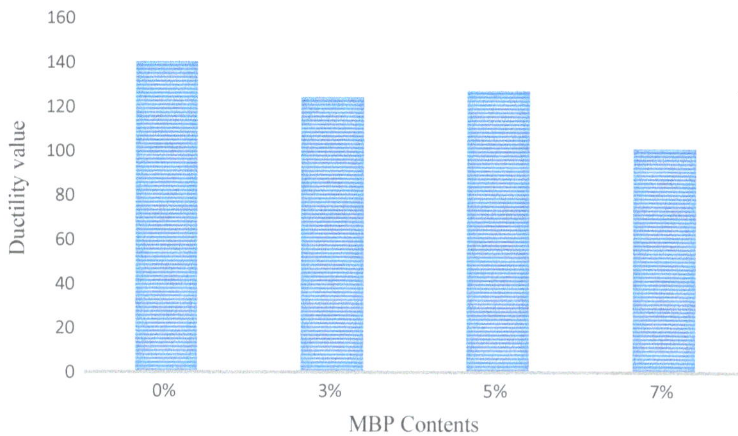

Fig. 3. Ductility of modified asphalt binders

C and MBP 7% was 49 °C. Additionally, it was perceived that the softening point value of the altered asphalt binder rose from 1 to 3 °C with an increase in the MPB percentage from 3 to 7%. However, an increase in MBP of 7° led to a drop in the softening point value. This is illustrated in Fig. 4.

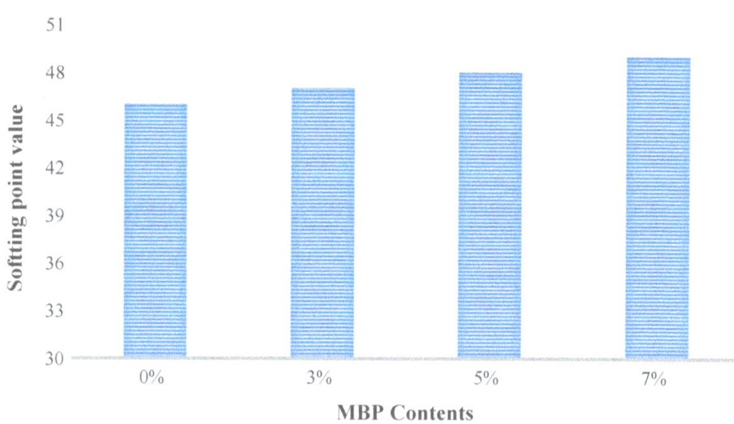

Fig. 4. Softening point of modified asphalt binders

3.2 Rheological Properties

The DSR was utilized to evaluate the rheological properties of asphalt binders under transitional to raised temperatures. Complex modulus (G*) and phase angle (δ) were the two parameters used for assessing the rheology (or viscoelastic properties) of the altered asphalt binders. G* represents the asphalt binder's sum resistance to irreversible deformation when frequently clipped. The higher the value of G*, the higher the asphalt binder's resistance level against permanent deformation. δ is defined as the phase angle flanked by stress and strain. It is strongly influenced by the temperature and frequency of traffic loading. This phase angle can also be utilized as a guide for evaluations on the elasticity and viscosity of asphalt binders. The lower the value of δ, the higher the elasticity level of the asphalt binder and its resistance against rutting deformation. In comparison to the asphalt binders modified by varying percentages of MBP, the base asphalt binder recorded the least complex modulus G* value at the temperatures tested. This is portrayed in Fig. 5. On the other hand, the phase angle value of the base binder was the highest in comparison to the other percentages of MBP modified asphalt binders. This is illustrated in Fig. 6. Thus, the MBP modified asphalt binder exhibited higher elastic properties and more resistance to rutting deformation than the base binder.

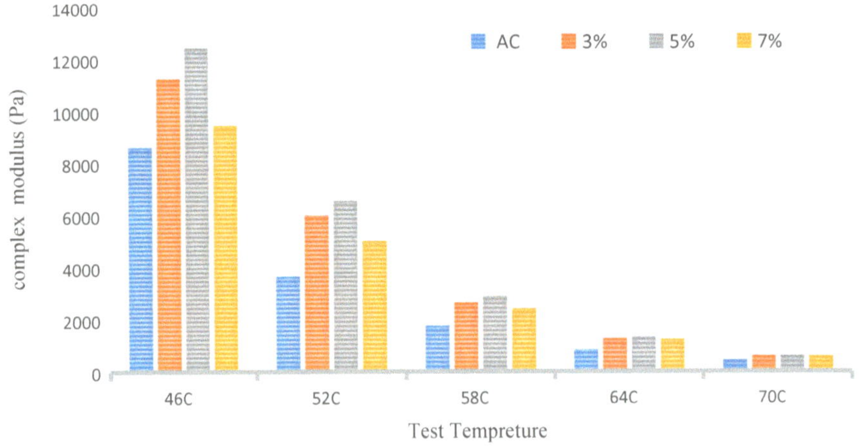

Fig. 5. The complex modulus (G*) against test temperatures

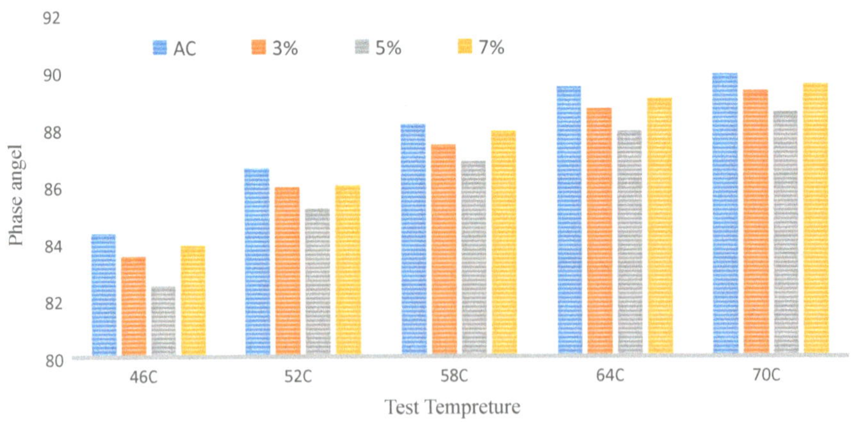

Fig. 6. The phase angle (δ) against test temperatures

4 Conclusion

The results attained from the physical properties tests (penetration, ductility and softening point) revealed that the stiffness of the asphalt binder was ~12% enhanced. The DSR results revealed a promotion in the (G*) and (δ) values at elevated temperatures. This is an indication that the viscoelasticity behaviour of the asphalt binders was improved. In other words, when compared to the base asphalt binder, the MBP 5% exhibited ~31% enhancing, which improved the resistance towards rutting deformation. As such, the test results confirm that the introduction of MBP into the asphalt binders served to enhance their properties, and thus their performance.

Acknowledgements. The authors would like to thank Prof. Madya Dr. Mohamad Yusri Bin Aman. We also acknowledge the contributions of Smart Driving Research Centre, Universiti Tun Hussein Onn Malaysia.

References

1. Al-tameemi, A. F., Wang, Y., Albayati, A.: Influence of hydrated lime on the properties and permanent deformation of the asphalt concrete, 9–11 (2015)
2. Soltani, M., Moghaddam, T.B., Karim, M.R., Baaj, H.: Analysis of fatigue properties of unmodified and polyethylene terephthalate modified asphalt mixtures using response surface methodology. Eng. Fail. Anal. **58**, 238–248 (2015). https://doi.org/10.1016/j.engfailanal.2015.09.005
3. Roberts, F.L., Kandhal, P.S., Brown, E.R., Lee, D.-Y., Kennedy, T.W.: Hot mix asphalt materials, mixture design and construction (1991)
4. Plunkert, B. P. a.: Auxite and Carbon (2001)
5. Bush, W.V.: History of bauxite in Arkansas, 8 (2007)
6. Patterson, S.H.: Bauxite reserves and potential aluminum resources of the world, 176 (1967)
7. Department, I., Survey, G.: Mineral Commodity Summaries: 2016. Government Printing Office (2016)
8. Ali, A.S.B., Idrus, M., Mohd, B., Musbah, A., Allam, A., Usman, N., Onn, H.: Influence of Parit Nipah peat (PNpt) on the properties of asphalt binder (2017)
9. Brown, E.R., Kandhal, P.S., Roberts, F.L., Kim, Y.R., Lee, D.-Y., Kennedy, T.W., et al.: Hot mix asphalt materials, mixture design, and construction. NAPA Research and Education Foundation, Lanham, MD (2009)
10. Idrus, M., Masirin, M., Musbah, A., Allam, A., Suliman, A., Ali, B.: Effect of Batu Pahat Soft Clay (BPSC) concentrations on the physical and rheological properties of asphalt binder. Sci. Technol. **25**, 101–108 (2017)

The Evaluation of the Performance of Nano Bauxite Powder (NBP) Modified Asphalt Binder

Hamza Mohamed Abdulhamid Elghatas[1(✉)], Mohamad Yusri Bin Aman[2], Shaban Ismael Albrka Ali[3], and Ahmed Suliman B. Ali[1]

[1] Faculty of Civil and Environmental Engineering, Universiti Tun Hussein Onn Malaysia, 86400 Parit Raja, Malaysia
alkatass@gmail.com
[2] Faculty Civil Environmental and Engineering, Smart Driving Research Centre, University Tun Hussein Onn, Parit Raja, Malaysia
mdyusri@uthm.edu.my
[3] Civil Engineering Department, Near East University, Near East Boulevard Nicosia, 99138 North Cyprus, Turkey
Shabarofking10@gmail.com

Abstract. For the first time, this study assessed the rheological and physical properties of asphalt binder combined with nano bauxite powder (NBP) derived from raw Malaysian bauxite. The physical properties were evaluated via conventional tests like softening point, penetration, rotational viscosity and ductility, while the impacts of NBP modified asphalt binders on rheological properties were assessed via dynamic shear rheometer (DSR). In addition, the morphology (nanoparticles size and shape) of the NBP was studied using transmission electron microscopy (TEM). The results show that using of NBP as a modifier increases the hardness of the asphalt binder and improves its physical properties. Compared with unmodified asphalt binder, there was $\sim 11\%$ enhancement in softening point and $\sim 16\%$ decrease in penetration of modified asphalt binders. In a similar manner, compared with base asphalt binder, there was increase in G* and G*/sin δ values for the modified asphalt binders and decrease in δ values, which suggests that there was an enhancement in the resistant property of the modified asphalt binders to permanent deformation (rutting). Therefore, to alter asphalt binder's properties, NBP is considered as an optimum additive.

Keywords: Nano bauxite powder · Physical proprieties · Dynamic shear rheometer and TEM

1 Introduction

With growing population worldwide, there has also been an increase in the number of vehicles globally, particularly in developed countries, which has resulted in a lot of pavement issues [1]. Asphalt pavement has to be able to withstand the loads in various climatic conditions, existing heavy loads and expected future loads for an acceptable

M. Awang and M. H. Isa (eds.), *The Advances in Civil Engineering Materials*,
Lecture Notes in Civil Engineering 19, https://doi.org/10.1007/978-981-13-2511-3_5

period. Therefore, to reduce the pavement distresses, the asphalt binder and asphalt mixes need to be enhanced. As early as the 1950s, the application of asphalt modifiers has been applied in the road structure industry to improve the performance of asphalt pavement [2]. Modifiers such as fillers, fibres, extenders polymer, oxidants, and nanocomposites materials had been applied for several years in asphalt binder to improve its properties.

Nanotechnology is microscopic particles of material, it is new materials that recently become popular and one of the large and important parts of research and development worldwide. It has been lighted on and used to control the problems in the design and construction of practical structures [3]. Nanotechnology is also defined as micro-scale fillers which can make improvements in the performance of the asphalt binder [4]. The asphalt pavement system can reap various benefits by employing nanotechnology as a modifier in asphalt mixture and asphalt binder. Some of the benefits include decrease in moisture damage, enhancement of asphalt properties during high and low temperatures, durability improvement of the asphalt mixture, energy saving and decrease in maintenance costs [3, 5–7].

In previous research studies, the performance of asphalt binder modified with different types of nanomaterials have been described. In addition, various papers have been published evaluating the performance of asphalt binder modified with polymers like EVA, SBR or SBS. However, in the literature, there is limited information regarding the NBP-modified asphalt binder's characteristics. Various contents of nanosilica were employed [8] to modify the asphalt binder. They found acceptable performance at high temperature with modified asphalt binders, suggesting it is better suited for high load traffic roads and high temperature regions. Additionally, [9] employed nano ZnO with various sizes of nanoparticles (average diameter of 2, 80 and 350 nm) to examine the performance of mixture and asphalt binder. The results of experimental evaluation show that the moisture susceptibility, flexural tensile strengths and low temperature bending strains of asphalt mixtures are improved with decreasing of ZnO particle size. Generally, an improvement in the performance grade leading to higher resistance to rutting deformation was obvious. The performance of nano bauxite powder-modified asphalt binder has been evaluated in this study considering modified asphalt binders' rheological and physical properties.

2 Experimental Design

2.1 Materials

We used asphalt binder 80/100 penetration grade as control asphalt binder, provided by Petronas Company (Malaysia). The raw material Malaysian bauxite was obtained from KK Best Metal Sdn. Bhd. Kuantan, Pahang, Malaysia, which is employed as a modifier to the asphalt binder.

2.2 Sample Preparation

2.2.1 Production of Nano Bauxite Powder (NBP)

The raw supplied bauxite was first ground into a specific powder size using LAAV machine and then, it sieved using a 75 μm sieve. Subsequently, the powder was dried at temperature 145 °C for getting rid of the moisture. After that, the bauxite powder was heated at 125 °C in an oven for 30 min. Thereafter, using a bowl mill machine the bauxite powder ground for 5, 10 and 15 h to measure the optimum grinding time.

2.2.2 Preparation of Modified Asphalt Binder

Into the base asphalt binder, addition of nano bauxite powder (NBP) was done in three different concentrations, namely 3, 5 and 7% NBP by weight of the asphalt binder. A high shear mixer at a speed of 3000 rpm was employed for mixing all modified asphalt binders. During the mixing process (approximately 60 min), temperature within the range of 160 ± 3 °C was maintained.

2.3 TEM

Transmission electron microscope (TEM, Hitachi HT7700) with high magnification and image resolution (1 nm) was employed to investigate the morphology (nanoparticles shape and size) of NBP.

2.4 Physical Properties

When evaluating the performance of the modified asphalt binder, the physical properties are critical parameters. The property changes of the asphalt binder modified with NBP, by comparing with the original asphalt binder, were assessed through physical tests such as ductility (ASTM D5), penetration (ASTM D5) and the softening point (ASTM D113).

2.5 Viscosity

The flow characteristics of the asphalt binder were assessed through the rotational viscosity test to get an initial idea if it can be handled and pumped when performing hot mixing. This study employed a Brookfield rotational viscometer to describe the modified and unmodified asphalt binder viscosity. Based on Superpave test parameters (Asphalt Institute 2007), the viscosity was measured by testing all samples with a spindle No. 27 that has a constant rotational speed of 20 rpm and maintained under temperatures of 135 and 165 °C.

2.6 Rheological Properties

The rheological properties of the unmodified and modified asphalt were determined by employing Dynamic Shear Rheometer (DSR). For all binders, a plate with a diameter of 25 mm and thickness of 1 mm was employed to characterise the viscous and elastic behaviour of asphalt binder at intermediate and high service temperatures conforming to AASHTO T315. Moreover, the impacts of employing NBP modified asphalt binders were investigated by measuring the phase angle (δ), complex modulus (G*) and rutting resistance parameter (G*/sin δ) of asphalt binder.

3 Results and Discussions

3.1 TEM

Figure 1 shows the TEM images of the NBP, with ellipsoidal nanoparticles while, the grain size distribution of the NBP during the grinding time which obtained from the TEM images is shown in Fig. 2. The agglomeration of nanoparticles in the micrograph attributed to the high surface energy and strong surface tension of ultrafine nanoparticles. Moreover, the fine particle size resulted in a large surface area, which enhanced the catalytic activity of the nanoparticle. The NBP grain sizes were 300–325, 125–150, and 10–20 nm for the grinding time durations of 5, 10, and 15 h, respectively. Grinding duration time of 15 h was selected as the optimum duration time to produce a nano bauxite powder with major nanoparticles size ranging from 10 to 20 nm.

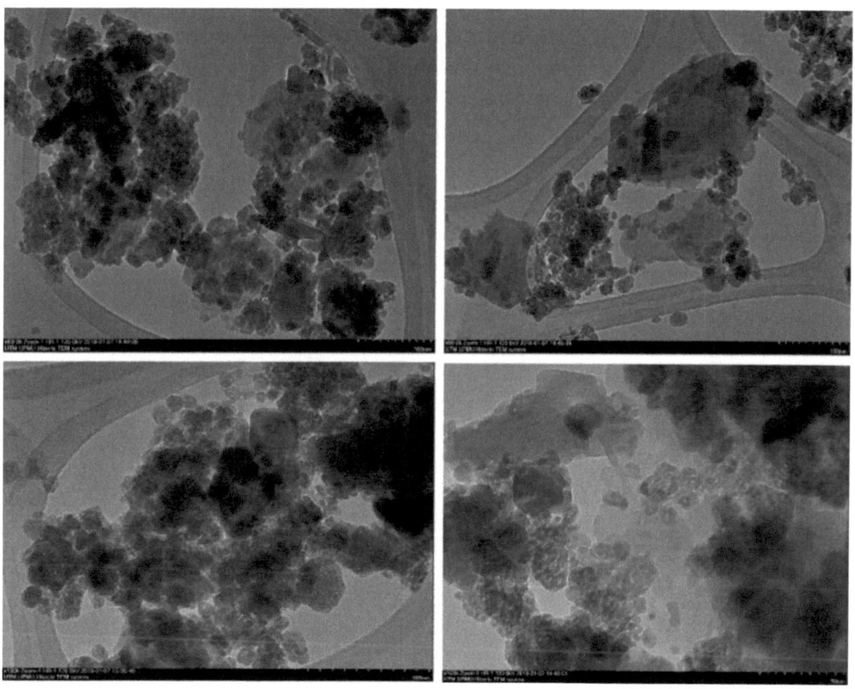

Fig. 1. Transmission electron microscopy (TEM) images of the nano bauxite powder (NBP)

3.2 Physical Properties

The impact of employing NBP as an additive along with the asphalt binder was recorded for the modified asphalt binder samples as physical properties. Figure 3 shows the relationship between the point values for ductility, penetration and softening and the NBP concentration. Rise in the NBP concentration was seen to cause a reduction in penetration values, which suggest that the modified asphalt binder has

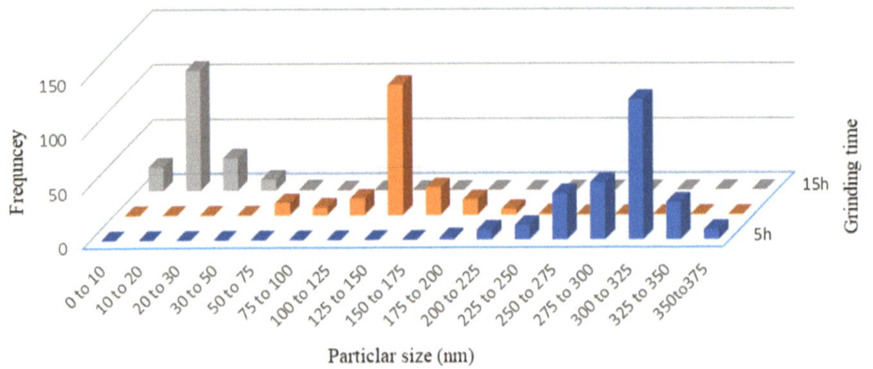

Fig. 2. Particular size distribution of the NBP during grinding time obtained from TEM images

become stiffer. Moreover, a reduction in the ductility values was seen due to increase in MBP concentration, as a result of the modified asphalt binder becoming stiffer. Moreover, the modified asphalt binders' softening point was improved because of increase in modified asphalt binder's hardness.

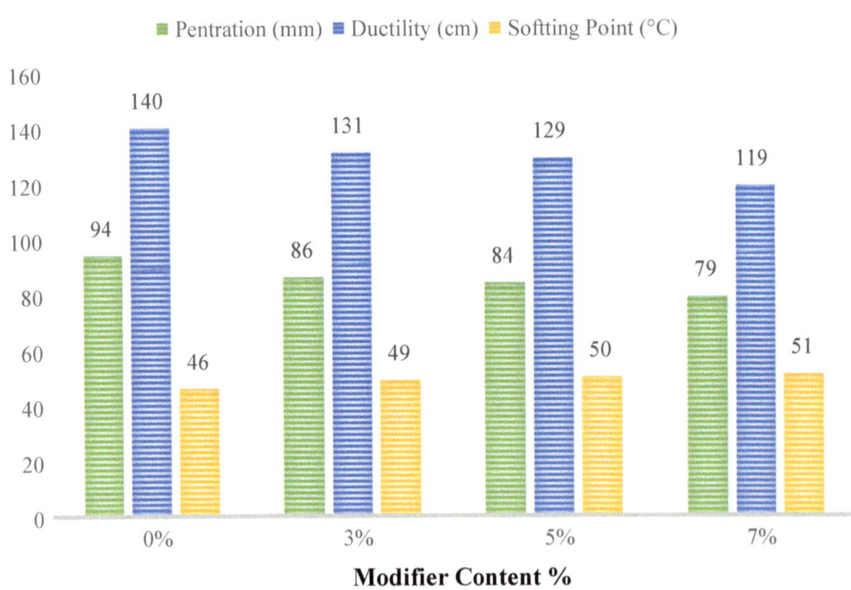

Fig. 3. Physical properties of base and modified asphalt binder

3.3 Viscosity

Figure 4 presents the rotational viscosity values for the base as well as modified asphalt binder. On adding NBP to the asphalt binder, the viscosity value was seen to increase at test temperatures (135 and 165 °C). The hardening effect of NBP led to increase in viscosity. Furthermore, the NBP layers' better dispersion in the base binder could also result in increase in viscosity value of the modified asphalt binder, which results in strengthening the bonding by restricting the asphalt flow. All these ultimately lead to increase in the hardness of the asphalt binder and enhancement of its physical properties.

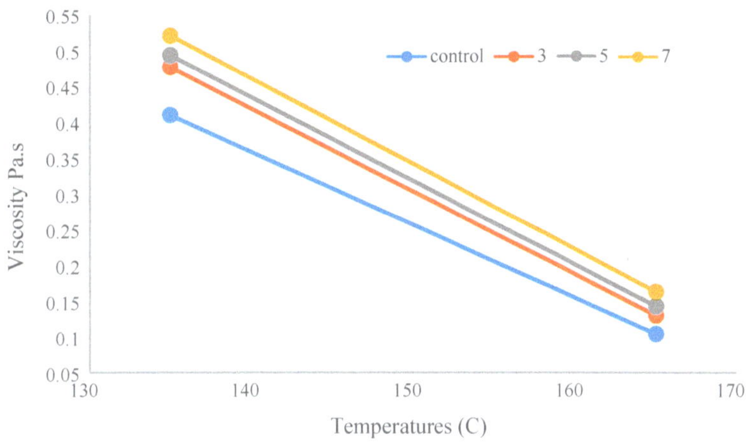

Fig. 4. Base and modified asphalt binder's viscosity

3.4 Rheological Properties

Complex modulus (G*) is also referred to as the asphalt binder's total resistance to rutting deformation, where the higher the value of G*, the greater will be the resistance to permanent deformation. Figure 5 shows the test temperature together with the complex modulus (G*) for the base and modified asphalt binder with NBP. The increase in G* value in the modified asphalt binder was observed. As seen in Fig. 5, it is obviously that 5% NBP has the highest value of complex modulus G* among the binders while the base binder has the lowest G* value, which means the modified asphalt binder with NBP has higher resistance to permanent deformation compared with the base asphalt binder.

The phase angle (δ) represents the angle between the strain and stress under the loading frequency, and it also relies on the temperature of the test. The phase angle (δ) is an indicator for defining viscous and elastic behaviours of the asphalt binder, in which the greater the value of δ, the higher the viscous behaviour it will display. Meanwhile, the low δ value represents higher elastic behaviour. Figure 6 presents the phase angle (δ) together with the test temperature for the unmodified and modified asphalt binders. The δ values for all modified asphalt concentrations reduced when compared with the base asphalt binder, which suggests that the addition of NBP improves the elastic behaviour. Also, as noted in Fig. 6, 5% NBP has the lowest value of δ compared to with the other percentages of the modified asphalt binder which means it has the highest elastic properties and is more resistant to rutting deformation.

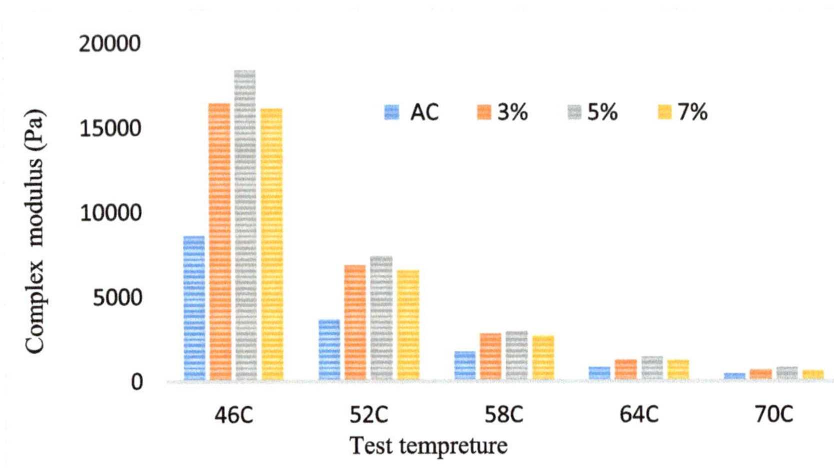

Fig. 5. Test temperature alongside the complex modulus (G*)

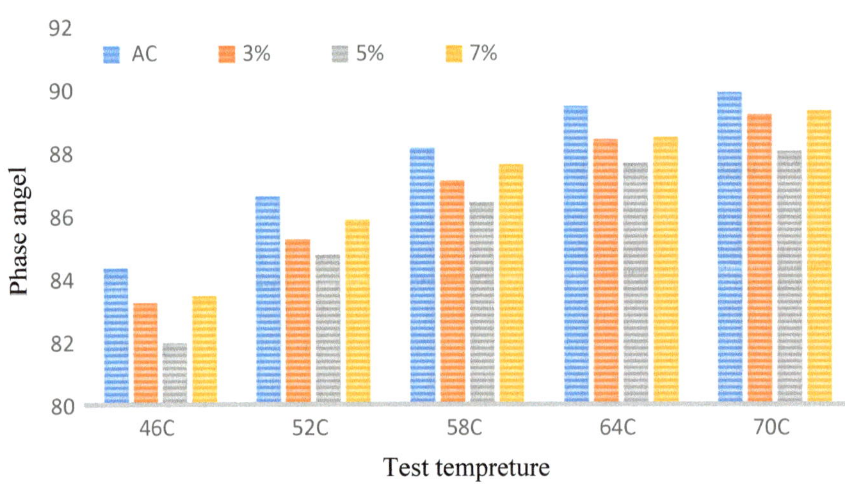

Fig. 6. Test temperature alongside the phase angle (δ)

Referring to the previous studies [10, 11], the rutting resistance of the asphalt binder can be characterised with the value of G*/sin δ at high temperatures. The superpave technique requires the minimum value for G*/sin δ to be 1 kPa for the rutting parameter of unaged samples [12]. Also, enhanced permanent deformation resistance pavement is confirmed with a higher value of G*/sin δ. As displayed in Fig. 7, the lowest value of G*/sin δ was associated with the base asphalt binder while the highest value of G*/sin δ was associated with the 5% NBP. Furthermore, for all the modified asphalt binder concentrations associated with NBP, an increase in G*/sin values was observed, which suggests enhanced permanent deformation resistance by employing NBP modified asphalt binder.

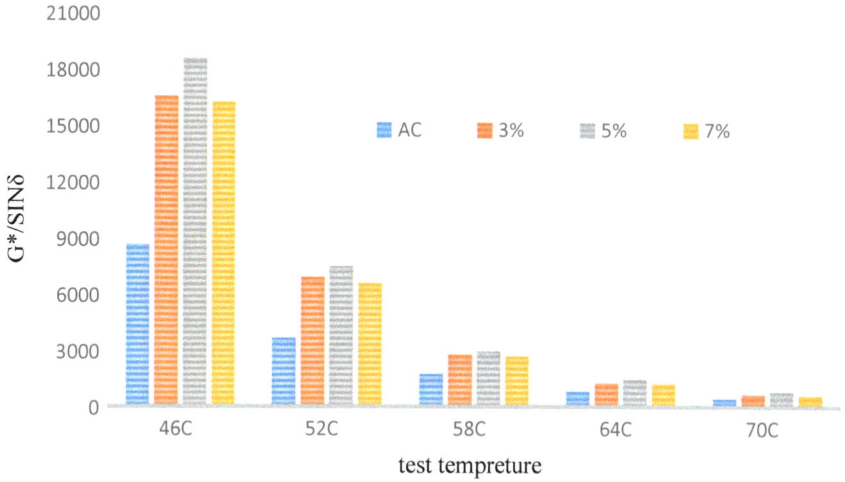

Fig. 7. Temperature's impact on permanent deformation (rutting)

4 Conclusion

The study included various tests for ductility, penetration and softening point, rotational viscosity and dynamic shear rheometer (DSR) to assess nano bauxite powder modified asphalt binder's performance. The following conclusions have been derived according to the obtained results:

i. NBP was successfully produced by grinding the raw bauxite for 15 h. TEM images showed the existence of ellipsoidal nanoparticles with a grain size ranging from 10 to 20 nm.
ii. As seen through conventional tests, employing NBP increased the stiffness of the modified binders, which also suggests enhancement of their temperature susceptibility, and improved softening point for all modified binders than with the base asphalt binder.
iii. There was an increase in viscosity values for all modified asphalt binders than with the base asphalt binder, which also improved the physical properties of the asphalt binder as well as increases its hardness.
iv. When compared with the base asphalt binder, there was an increase in the values of G* and G*/sin δ for the modified asphalt binders along with a decrease in δ values, which suggests enhanced resistance to permanent deformation (rutting).

In general, enhancement in the rheological and physical properties of asphalt binder was seen when the NBP-modified asphalt binder was employed. Further, the optimum percentage that is considered for the modifier can therefore be 5% NBP.

References

1. Ismael Albrka, S., Ismail, A., Yahia, H.A.M., Azizul Ladin, M.: Application of transyt-7f on signalized road junction networks in Shah Alam and Petaling Jaya. Jurnal Teknologi (Sciences and Engineering), **69**(2), 59–64 (2014). https://doi.org/10.11113/jt.v69.3108
2. Roberts, F.L., Kandhal, P.S., Brown, E.R., Lee, D.-Y., Kennedy, T.W.: Hot mix asphalt materials, mixture design and construction (1991)
3. Yang, J., Tighe, S.:. A review of advances of nanotechnology in asphalt mixtures. Procedia—Social Behav. Sci. **96**(Cictp), 1269–1276 (2013). https://doi.org/10.1016/j.sbspro.2013.08.144
4. Albrka Ali, S.I., Ismail, A., Md. Yusoff, N.I., Abdul Hassan, N., Ibrahim, A.N.H.:. Characterization of the performance of aluminium oxide nanoparticles modified asphalt binder. Jurnal Teknologi (Science & Engineering) **78**(4), 91–96 (2016)
5. Hainin, M.R., Matori, M.Y., Akin, O.E.: Evaluation of factors influencing strength of foamed bitumen stabilised mix. Jurnal Teknologi **70**(4), 111–119 (2014). https://doi.org/10.11113/jt.v70.3499
6. Hussein, A.A., Jaya, R.P., Abdul Hassan, N., Yaacob, H., Huseien, G.F., Ibrahim, M.H.W.: Performance of nanoceramic powder on the chemical and physical properties of bitumen. Constr. Build. Mater **156**, 496–505 (2017). https://doi.org/10.1016/j.conbuildmat.2017.09.014
7. Mubaraki, M., Ali, S.I.A., Ismail, A., Yusoff, N.I.M.: Rheological evaluation of asphalt cements modified with ASA polymer and Al_2O_3 nanoparticles. Procedia Engineering **143** (Ictg), 1276–1284 (2016). https://doi.org/10.1016/j.proeng.2016.06.135
8. Abdullah, M., Zamhari, K., Nayan, N., Hainin, M., Hermadi, M.: Physical properties and storage stability of asphalt binder modified with nanoclay and warm asphalt additives. World J. Eng. **9**(2), 155–160 (2012)
9. Zhang, H., Gao, Y., Guo, G., Zhao, B., Yu, J.: Effects of ZnO particle size on properties of asphalt and asphalt mixture. Constr. Build. Mater. **159**, 578–586 (2018)
10. Ali, S.I.A., Ismail, A., Yusoff, N.I.M., Karim, M.R., Al-Mansob, R.A., Alhamali, D.I.: Physical and rheological properties of acrylate-styrene-acrylonitrile modified asphalt cement. Constr. Build. Mater. **93**, 326–334 (2015). https://doi.org/10.1016/j.conbuildmat.2015.05.016
11. Idrus, M., Masirin, M., Musbah, A., Allam, A., Suliman, A., Ali, B.: SCIENCE & TECHNOLOGY Effect of Batu Pahat Soft Clay (BPSC) Concentrations on the Physical and Rheological Properties of Asphalt Binder, 25, 101–108 (2017)
12. Khadivar, A., Kavussi, A.: Rheological characteristics of SBR and NR polymer modified bitumen emulsions at average pavement temperatures. Constr. Build. Mater. **47**, 1099–1105 (2013)

Behavior of Laminated-Timber Slab Using Mechanical Connector

A. P. Yosafat(✉), M. Roi, H. S. Jean, I. Y. Sarah, and T. L. Chris

Department of Civil Engineering, Universitas Kristen Maranatha, Bandung, Indonesia
yosafat.ap@gmail.com, yosafat.ap@eng.maranatha.edu

Abstract. The behavior of the floor slab supports the function of the slab. In high-rise residential buildings, the behavior that needs to be known is due to the presence of static load and vibration load. Behavior due to static load is the flexural behavior of the floor slab due to the loads that work on it is the gravitational load. This study discussed the bending behavior of laminated timber floor slab due to static load. The scope of the research is the wood used is Mersawa wood (*Anisoptera* spp.), The test used three point loading test in accordance with ASTM D143, the behavior discussed is flexural strength, the laminate floor slab is arranged horizontally with the mechanical connector that is bolt with 10 mm diameter. The test shows that the average bending capacity (proportional limit load) is 25.65 kN. The failure mode occur is simple tension. The result of the test indicated that, in general the proportional limit load is higher than service load, so the floor slab can be used for residential building, such as multistory buildings. The deflection of wooden floor slabs due to static and vibration loads is smaller than the permit deflection limit. These results indicate that the slab has sufficient stiffness and has adequate comfort to the vibration.

Keywords: Slab · Timber · Laminated · Bolt · Flexural

1 Introduction

Indonesia is a tropical country which is rich in natural resources, such as timber. Timber is one of the building materials that are widely used as supporting the structure of the building. In Indonesia timber as a construction material is still widely used, among others, used for midrise buildings, houses, bridges, train bearings, and others. Especially in building buildings, timber construction is widely used to make trusses, beams, columns, and slabs. In Indonesia, traditional wooden houses such as stage houses mostly use wooden construction [8]. Along with the development of time, timber construction is still widely used in modern buildings both regular houses and multi-story buildings. One of the many wooden structure components is the floor slab. The floor slab is the structure that first receives the load, both dead and live loads which then channel to another frame structure system (Fig. 1).

M. Awang and M. H. Isa (eds.), *The Advances in Civil Engineering Materials*,
Lecture Notes in Civil Engineering 19, https://doi.org/10.1007/978-981-13-2511-3_6

(a) Beam and Colum System. (b) Beam setup for house-floor.

Fig. 1. Minangkabau traditional wooden house [8]

Timber is increasingly difficult to find especially solid wood with large dimensions and long spans due to illegal logging that occurs more and more, especially in Indonesia. The average available timber has a small diameter and short span whereas in the construction of buildings used timber that has a large diameter and long span. This scarcity of solid timber has resulted in consultants seeking solutions to this problem. One solution is the method of timber engineering. One example of the method of timber engineering is laminated timber. Laminated timber is processed wood consisting of several layers of wood arranged in such a way as to form a unity of lamina. The use of laminated timber is more advantageous from several sides. The advantages include: stronger power three times or equivalent solid sawn timber, expanding very small wood shrinkage, and varying wood dimensions. The use of laminated timber is highly recommended for earthquake prone buildings. So laminated timber is preferred by consumers.

In addition, another problem faced is the application of laminated timber to the structural system in an innovative and easy to implement in the field. Laminated timber with a mechanical system (bolts) can be a solution to the problem. One example of applying laminated timber to the structural system is the floor slab. Material selection for floor slabs is an important thing to consider considering the floor slab is the structure that first receives the load which then channeled to another truss system.

The aim of this research is to study behavior of the bending behavior of laminated timber floor slab due to static load, which are strength both at proportional load and ultimate limit load, stiffness or ductility of floor slab. The scope of the research is the wood used, which is Mersawa wood (*Anisoptera* spp.), The test used is three point loading test in accordance with ASTM D143 [2], the behavior discussed is flexural strength, the laminate floor slab is arranged horizontally with the mechanical connector that is bolt with 10 mm diameter and 200 mm spaces between bolts. Specimens used for the study amounted to 3 (three) specimens.

2 Basic Theory

2.1 Bending Strength, Shear Strength, and Deformation

Basic theory used for design are bending strength, shear strength, and deformation in accordance with SNI 7973:2013 [3]. The normal stress analysis of bending section beams due to bending is used to calculate the flexural stress distribution at the cross section in the middle of the span. From the calculation result that is the bending stress on the tensile part, it can be known the magnitude of tensile force [5]. Flexural nominal strength (for beam with rectangular cross section) in accordance with SNI 7973:2013 [3] which is the design method of LRFD (Load Resistance Factor Design),

$$f_b \leq F'_b \tag{1}$$

where f_b and $F_b{}'$ are,

$$f_b = \frac{M.y}{I_x} \tag{2}$$

$$F'_b = F_b.C_M.C_t.C_L.C_F.C_{fu}.C_r.C_i.K_F.\phi_b.\lambda \tag{3}$$

where f_b is normal stress due to flexural, M is flexural moment, y is distance from centroid axis to edge of beam, $F_b{}'$ is adjusted bending design, F_b is reference bending design (flexural strength parallel to the grain), C_M is wet service factor, C_t is temperature factor, C_L is stability correction factor, C_F is size factor for sawn lumber, C_{fu} is flat use factor, C_r is repetitive member factor for dimension lumber, C_i is incising factor for dimension lumber, K_F is format conversion factor, ϕ_b is resistance factor for bending, λ is time effect factor.

The design rules of shear strength used by the design method of LRFD (Load Resistance Factor Design) are reviewed from SNI 7973: 2013 [3]. The following equations of parallel shear strength requirements of parallel fiber on solid wood are,

$$f_v \leq F'_v \tag{4}$$

where f_v and $F_v{}'$ are,

$$f_v = \frac{V.Q}{(I_x.b)} \tag{5}$$

$$F'_v = F_v.C_M.C_t.C_i.K_F.\phi_v.\lambda \tag{6}$$

where f_v is actual shear stress parallel to the grain, $F_v{}'$ is adjusted shear stress parallel to the grain, V is shear force, F_v is reference shear stress parallel to the grain, Q is static moment of cross section, I_x is inertia moment of cross section, b and d are width and height of beam, and ϕ_v is resistance factor for shear.

The design of the bending moment force used in the LRFD design method is reviewed from SNI 7973: 2013. In addition, there are several provisions concerning the beam stability factor. The slenderness ratio (R_B) for the bending structure component, shall meet the requirements,

$$R_B = \left(\frac{l_e.d}{b^2}\right)^{0.5} \leq 50 \tag{7}$$

where l_e is effective length of beam and l_u is actual length of beam. To review the requirements analysis of structural deflection requirements,

$$\Delta' \leq \Delta_{allowed} \tag{8}$$

where Δ' is deflection of beam due to loads.

2.2 Methods for Determine Proportional and Ultimate Loads

Knowledge of the method of determining the proportional point and the ultimate limit loads is very important. There are several methods that can be used to determine this point, especially for wood materials [6] i.e. EEEP and Yasumura and Kawai methods.

The EEEP (Equivalent Energy Elastic-Plastic Curve) method is a commonly used method of steel material and often wood material, i.e. curve modeling into a perfect elastic-plastic behavior model (Fig. 2a). The extent of the empirical test result curve is assumed to be equal to the area of the bilinier curve. In the Yasumura and Kawai methods, initial rigidity (straight line) is calculated between the range of 10–40% of the maximum load. Next is defined a straight line between two points where the value of 40 and 90% of the maximum load. The melting point is determined from the meeting of the two lines (Fig. 2b). Figure 3 shows an example of load determination under proportional load (P_y) load conditions and loads on ultimate boundary conditions (P_u). Thus it can also be determined by the deformation in term of proportional (d_y) and ultimate (d_u) limits.

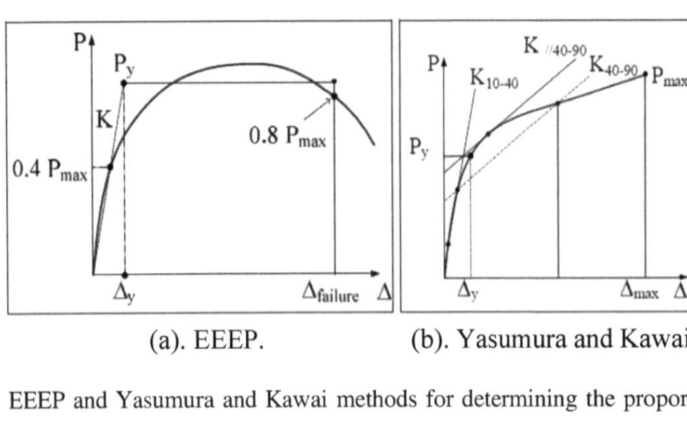

(a). EEEP. (b). Yasumura and Kawai.

Fig. 2. EEEP and Yasumura and Kawai methods for determining the proportional [6]

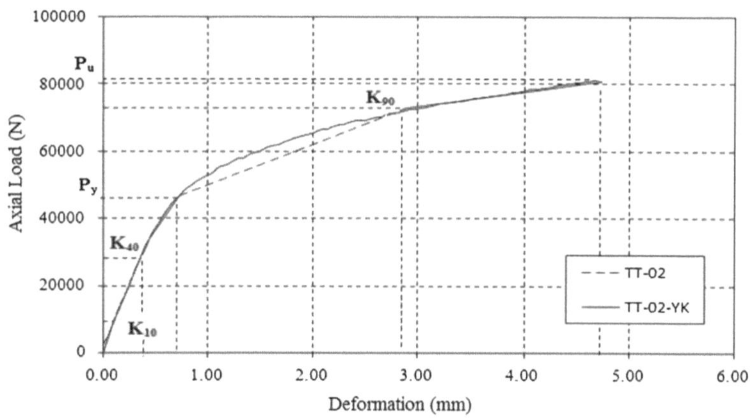

Fig. 3. Example determines the proportional and ultimate limit loads using Yasumura and Kawai method [7]

2.3 Effect of Vibration

Floor structure planning can not be separated from the influence of the vibration, both temporary with short duration and long duration. The vibration serviceability condition of the wood floor slab structure is related to the size of the floor span, the longer the span the impact of the vibration will be greater. Some wood regulations govern the requirements regarding the conditions of comfort limits to vibrations. In Indonesian SNI wood regulations SNI 7973: 2013 [3], specifically has not regulated this matter. From the literary sources of Weckendorf et al. [9], it can be learned that attenuation is an important characteristic of the vibration response of the structural system. Damping can convert kinetic energy into heat, reducing the amplitude of the force or free dynamic movement. In the timber regulation of Eurocode 5 [4] there are design criteria related to vibration in the wood floor slab structure system. Criteria designs include vibrations generated by mechanical or electronic machines placed on floor systems, permissible serviceability requirements, and damping ratio estimation calculations. The Eurocode 5 regulation recommends a 1% damping ratio. This regulation recommends for one segment system structure, the fundamental capital frequency is 11.5 Hz. The existence of vibration frequency resulted in the existence of mass capital, according to the rules of Design Guide for Floor Vibrations [1] then calculation of mass capital can be calculated by equation as follows,

$$f_1 = \frac{18}{\delta} \tag{9}$$

$$M_{mod} = M.\left[\frac{\delta_x^2 + \delta_y^2}{2.\delta^2} + \frac{8}{\pi^2}\frac{\delta_x + \delta_y}{\delta^2}\right] \tag{10}$$

with f_1 being the first eigen frequency and δ is the deflection due to the gravitational load and M_{mod} is the mass capital. M_{mod} is further used as an additional mass due to vibration. In this proposal, the deflection discussed is a deflection in the direction of gravity (y-axis).

3 Case Study and Discussion

In this research, the specimens that were used (see Fig. 4), previously had preliminary analysis, and from the results of the analysis that has been done, it can be concluded that the mechanical laminate flooring slabs resulting from this research are mechanized laminated timber flooring slabs (*b*) 450 mm × (*d*) 60 mm, 5 lamina, 90 mm thick laminate, vertical laminate system. The length of each beam is 1050 mm. Total length of beam span with connection 1200 mm. Mechanical lamination tool is a 10 mm bolt diameter, and space distance per 20 cm. Specification of wood material and connection material used in this research are Mersawa wooden rod (*Anisoptera* spp.) with cross-sectional size (after 4 shades) 90 mm × 60 mm, used to make wooden floor slabs. The length of the rod is 1200 mm. The number of logs required to make the specimen is 5 (five) laminae. Steel bolt with 10 mm diameter with flexural yield strength Fyb equal to 310 MPa (As per guidance of SNI 7973: 2013). Steel ring with an outer diameter of 26, 13 mm inner diameter, and 0.95 mm ring thickness. Figure 4 show the setup of the specimens on Universal Testing Machine.

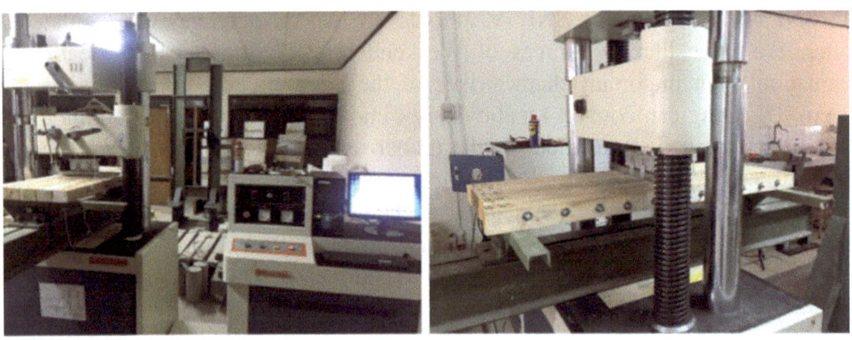

(a). Universal Testing Machine (UTM). (b). Specimen placed at UTM.

Fig. 4. Setup of the specimens on universal testing machine

Figure 5 shows the results obtained from tests. Failure of the specimens after reach the ultimate loads. The failure mode occur is simple tension. Table 1 show the results of Proportional and Ultimate Limit, both for Loads and Displacements that obtained from tests for all 3 (three) specimens. Methods used are Yasumura and Kawai Methods [6].

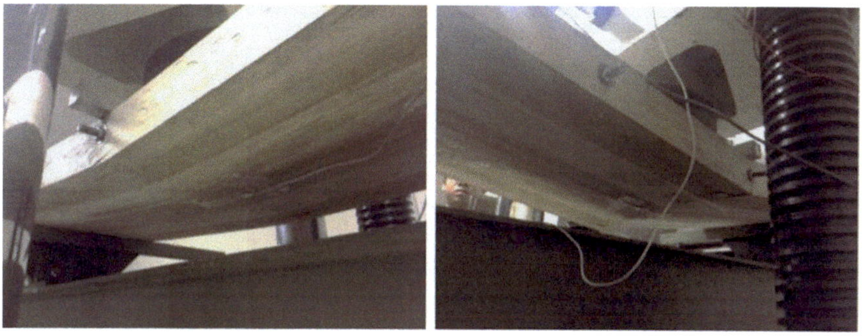

(a) Simple tension failure of S20-B01. (b) Simple tension failure of S20-B02.

Fig. 5. Failure of the specimens after reach the ultimate loads

Table 1. Proportional and ultimate limit load and displacement obtained from tests

Specimen	P_y (N)	D_y (mm)	P_u (N)	D_u (mm)	μ
S20-B01	31.21	6.78	41.11	12.54	1.85
S20-B02	21.34	8.23	26.05	12.40	1.51
S20-B03	24.39	10.73	29.18	23.69	2.21
Average	25.65	8.58	32.11	16.21	1.85

Figure 6 show the results obtained from tests which are load-displacement curves of bending tests. The laminated wood floor slab is designed to withstand maximum load (P) of 31.51 kN. This load is the maximum load limit based on the reference strength of sawn timber E18 according to the reference of SNI 7973: 2013. These results indicate that the beam has sufficient strength according to SNI regulations. The deflection of the wooden floor slab (Δ) is smaller than the permit deflection limit (allowed). These results indicate that the beams have sufficient stiffness according to SNI regulations. Additional mass calculations due to the effect of vibration using Eqs. 9 and 10,

$$f_1 = \frac{18}{0.35} = 51.43\,\text{Hz}$$

$$M_{mod} = 293.46\left[\frac{0^2 + 0.35^2}{2 \times 0.35^2} + 0\right] = 146.73\,\text{kg}$$

Therefore, due to the additional mass due to vibration, the effect on the amount of deflection. From the analysis of the structure of the floor slab to the effect of gravity and vibration loads. From result of modeling SAP2000, got deflection result of floor slab equal to 0.316 mm while the limit is 3.5 mm, so the deflection of the floor slab $\Delta < \Delta_{allowed}$ then the slab meets the comfort requirements of the vibration due to the load.

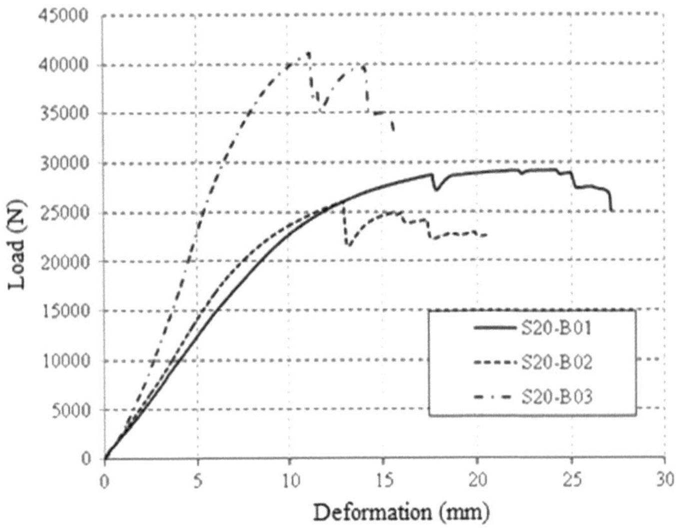

Fig. 6. Results obtained from tests

4 Conclusion

The test shows that the average bending capacity (proportional limit load) is 25.65 kN. The failure mode occur is simple tension. The result of the test indicated that, in general the proportional limit load is higher than service load, so the floor slab can be used for residential building, such as multi-story building. The deflection of wooden floor slabs (Δ) is smaller than the permit deflection limit and indicate that the slab has sufficient stiffness according to SNI regulations. Deflection due to vibration of wooden floor slabs (Δ) is smaller than the permit deflection limit and indicate that the slab has adequate comfort to the vibration of the load.

References

1. Arcelor Mittal.: Design Guide for Floor Vibrations, Arcelor Mittal Europe—Long products Sections and Merchant Bars. Arcelor Mittal (1996)
2. American Standard Testing and Material.: ASTM D143, American Standard Testing and Material (2008)
3. Badan Standarisasi Nasional.: Spesifikasi Desain Untuk Konstruksi Kayu (SNI 7973:2013). BSN, Jakarta (2013). (in Indonesian)
4. CEN.: Comite´ Europe´en de Normalisation (CEN) Eurocode 5-design of timber structures, part 1–1: general—common rules and rules for buildings. EN 1995-1-1:2004 (E), English version. British Standards Institution, London, UK (2004)
5. Gere, J.: Mechanics of Materials. Brooks/Cole (2009)
6. Munoz, W., Mohammad, M., Salenikovich, A., Quenneville, P.: Determination of yield point and ductility of timber assemblies. In: Search for a Harmonized Approach. Engineered Wood Products Association (2010)

7. Pranata, Y.A., Suryoatmono, B: Kekuatan Tekan Sejajar Serat dan Tegak Lurus Serat Kayu Ulin (Eusideroxylon Zwageri), Civil Eng. J. Bandung Inst. Technol. Indonesia. **21**(1) (2014, April). (in Indonesian)
8. Pranata, Y.A., Tobing, H.L.: Non-destructive testing to obtain the dynamic elastic modulus of the existing Minangkabau wooden house. In: International Conference on Technology, Innovation, Society (ICTIS) 2016, ISBN 978-602-70570-4-3, ITP Press (2016)
9. Weckendorf, J., Toratti, T., Smith, I., Tannert, T.: Vibration serviceability performance of timber floors. Euro. J. Wood Prod. **74**(3), 353–367 (2016)

The Importance of Culture in Disaster Management in Malaysia

Hayder Jawad Shakir Haraty(✉) and Nangkula Utaberta

Department of Architecture, Faculty of Design and Architecture, Universiti Putra Malaysia, 43400 UPM Serdang, Malaysia
hayderharaty@gmail.com

Abstract. The number of natural hazards, defined as threats or risks of naturally occurring disasters, has increased noticeably in recent years. Governments on the local and international level have made enormous efforts to develop systematic techniques and form agencies to deal with the occurrence of disasters. Learning from sequences of past crises could significantly enhance the awareness, preparedness, and response of people toward natural disasters. Moreover, understanding human responsibility is key, as mistreatment of the environment and poor planning in rapid community development impacts and increases the occurrence of natural disasters. This paper discusses the issue of culture in disaster management through an intensive analysis of the relevant literature. Despite the framework of the Malaysian government to increase public awareness of natural hazards leading to disasters, the Malaysian community still has little knowledge of risk reduction. Research reveals that the integration of culture in disaster management has proven to be significant in reducing the number of disaster-related deaths.

1 Introduction

Disasters can be defined as sudden occurrences in nature causing loss of life and property as well as environmental effects that impact citizens' daily activities, and such disasters are often believed to be unavoidable [1]. However, natural disasters are also a social phenomenon, and there is human agency involved in the prevention and management of disasters. Weakness of society awareness is an important factor in the occurrence of natural disasters [2]. The first issue was the miscommunication between the monitoring Centre's departments regarding the eruption of the Indian Ocean tsunami. The Pacific Tsunami Monitoring Centre (PTMC) in Honolulu, Hawaii, detected two signals; the first came 18 min after the eruption of the tsunami which was believed to be a false alarm. The second one was 65 min later which caused the Indian Ocean tsunami tragedy. It was not required for government officials to communicate with the Indian Ocean region as there was no such contract or commitment in place. Furthermore, it would not have made a difference if the disaster had been announced as the region had no effective civil defense mechanism to distribute information to people. This circumstance was unprecedented.

The International Federation of Red Cross and Red Crescent Societies (IFRCRCS) define disaster management as the organization and oversight of resources and

M. Awang and M. H. Isa (eds.), *The Advances in Civil Engineering Materials*,
Lecture Notes in Civil Engineering 19, https://doi.org/10.1007/978-981-13-2511-3_7

responsibilities for dealing with all humanitarian aspects of emergencies, particularly preparedness, response, and recovery, in order to lessen the impact of disasters [3]. The number of natural hazards has rapidly increased in recent years due to several factors, such as climate change, new patterns of marginalization, and population growth. An increased number of disasters has resulted in thousands of deaths, while the cost of these disasters has reached over billions of dollars [3, 4]. The Center for Research on the Epidemiology of Disasters (CRED) estimated the number of deaths by natural disasters at 235,000, with 2.14 million people affected and over $190 billion spent in 2008 alone [5]. Many strategies have been applied for dealing with small-, medium-, and large-scale disasters to help vulnerable people worldwide. It is essential to promote a culture of safety and preparedness among people to reduce the risk of disaster.

Based on recent natural disasters incidents, undoubtedly, there is no country that has not encountered the threat of disaster; however, there are several levels of disaster that differ from country to country. Preparedness and prevention may not be adequate sometimes because the risk of disaster is various and often based on geographic location. The Asian-Pacific region, suffered from frequent natural hazards and disasters from 1995 to 2004. China, Indonesia, Iran, and Pakistan have a high risk of earthquakes; small islands like the Maldives have risks from the sea; Bangladesh, parts of China, and India have suffered from frequent floods [6].

Furthermore, risk reduction and disaster management have no standardized cycle. Different agencies have different roles depending on their objectives. However, disaster management is carried out by a framework of general principles: Mitigation, Risk reduction, Prevention Preparedness, Response, and Recovery.

There are many key players in the disaster warning and management process [7]: communities, local governments, national governments, regional institutions and organizations, international bodies, non-governmental organizations, the private sector, the media, and the scientific community. Research shows how different agencies, governments, and communities have dealt with recent disastrous events, and specific examples will be discussed further in this article.

2 The Role of Government and Agencies

The impact and aftermath of the Indian Ocean tsunami in December 2004, which caused over 350 thousand deaths (with over 50 thousand missing), focusing on economic issues [8]. Sri Lanka and Indonesia were among the countries that suffered the most from this tragedy. The research also examines the role of the public, international agencies, and organizations in disaster relief after the tsunami. While the Indian Ocean tsunami is considered unique and one of the largest natural disasters in recent history, it has been valuable for the study of disaster management, as it gained global attention and numerous countries sent aid. The vast impact of the tsunami created difficulties for agencies and international organizations in offering logistical help; therefore, establishing a disaster monitoring system that increased public awareness on a global level was necessary. There were various roles needed to reduce workload and oversee preparations. The major role of the agencies was to avoid duplication of tasks and setting up procedures for translating aid pledges into actual aid flows; individual donors played their part in disaster relief by using informal private channels to gain funds and to avoid untoward effects of massive aid inflows.

In response to the disaster, $6.3 billion was pledged by 2005 from 78 countries and 30 organizations and individuals [9]; another $1.6 billion was received from private contributions, and in some donor countries, such as the United Kingdom, the United States, and Italy, private contributions exceeded the sizable government contributions by a wide margin [8]. Table 1 provides statistics from the 2004 tsunami and highlights the various types of donors who contributed aid for recovery. Contributing organizations are listed: FAO (Food and Agriculture Organization), UNDP (United Nations Development Programme), UNEP (United Nations Environment Programme), American Red Cross, Arab Gulf Fund, Asian Development Bank, Disasters Emergency Committee (UK), European Commission, European Commission, EuropeAid Co-operation Office, European Commission Humanitarian Aid Office, ROLEX, UNICEF NATCOMs, United Nations Fund for International Partnership, and Private sectors. Commitments for aid were made after the tsunami occurred, and most aid was sent to multilateral relief organizations, such as UNICEF, WHO, ILO, WFO, Red Cross, and various NGOs.

A study from [10] highlights the agencies' joint efforts to indicate on a map where the tsunami would most likely hit in Indonesia and their approach to a systematic and comprehensive monitoring system. Those agencies are the Meteorology and Geophysics Agency, Volcano and Geological Disaster Mitigation Directorate, The Ministry of Energy and Mineral Resources, and Center of Geological Research and Development. However, due to the fact that there was not enough funding, the program failed; even with the existence of such a system, it is doubtful that it would have made a difference in the 2004 tsunami as it was massive and reached land in such a short time

Table 1. The tsunami devastation summary (February 2005)

Country	Areas affected	Damage	Displaced	Deaths	Missing people
India	2200 km of costal land; 300 m to 3 km island, and 3 million people	897 villages, 157,393 dwelling units, 11,827 ha of cropped area, US$1.56 bn assets	647,556	10,872	5551
Indonesia	Aceh, 14 out of 21 districts; 1 million people	172 sub-districts, 1550 villages, and 21,659 houses	811,409	166,760	127,749
Malaysia	North West states of Penang and Kedah		8000	68	6
Maldives	20 atolls	100,000 people	10,578	82	26
Myanmar	23 villages	592 houses of 17 villages	2591	At least 2500	–
Sri Lanka	1720 km of coastal land; 300 m to 3 km inland, 103 families.	78,529 fully damaged houses; 41,097 partly damaged houses	502,668	30,959	5644
Thailand	6 provinces on the west coast	6.85 million baht have been provided to assist victims	–	5392	3100
Somalia	Puntland region worst hit, 650 km coast line	600 families have lost properties. 2600 fishing boats destroyed	4000	At least 150	N/A

Source www.economist.com

(15 min). However, Disaster management in Indonesia has developed alternative strategies to respond to the natural disaster. Disaster management involves the government's response to a massive natural disaster in three phases [8]: emergency and rescue operations, rehabilitation of basic socio-economic infrastructure, law, and order, and rebuilding of the economy and governmental system.

A study published [11] discusses the 2004 tsunami, its impact on Thailand, and the role of disaster management. The tragedy left 5395 dead and 2822 missing in Thailand. The research promotes strengthening disaster mitigation and preparedness in order to enhance the community's resilience in such events. It is undeniable that earthquakes and ensuing tsunamis cannot be prevented; however, it is possible to reduce the destructive aftermath and keep the number of deaths to a minimum through an integrated approach that includes prevention, preparedness, response, and recovery. In Thailand, however, disaster management mostly focused on emergency response and post-impact recovery, and after the 2004 tsunami, such approaches proved to be insufficient to deal with this kind of disaster. The aforementioned research shows that more attention should be given to awareness and preparedness among a community. The Thai government, in response to the 2004 tsunami, established the National Disaster Warning Center to receive, monitor, process, and relay critical information to the public before and during natural disasters [12]. The program included several actions to reduce current and future hazards. One of these was to develop educational programs. The aim was to educate people and students on natural hazards and grow a culture of understanding about surviving such events.

3 Disaster Management in Malaysia

The location of Malaysia near the equator and the climate influenced by the tropical belt cause landslides and floods every year [4]. Table 2 shows the number of disasters that hit Malaysia from 1993 to 2004. Several researchers [13] have indicated that major factors causing serious incidents in Malaysia are the poor understanding of geo-hazard phenomena, rapid development, and environmental degradation. Disaster preparedness is a concern on the national agenda [1].

Moreover, the research of [14] explains the role of management systems in disaster management. Their research was an early warning in Malaysia to understand recovery management, disaster response, and collaboration among organizations and agencies for humanitarian assistance in disaster relief. Furthermore, the researchers state that the location of Malaysian involves several kinds of disasters, such as landslides and floods; however, increasing a culture of awareness and prevention might help in managing future incidents and increase safety in the community. There is an urgent need to summarize, organize, and share knowledge of disasters in Malaysia [14].

Information and Communication Technologies (ICT) is a system used in Malaysia to increase the level of preparedness among Malaysians in the occurrence of a natural disaster [6]. The system has proven its effectiveness as it gathers important information regarding natural disasters and contributes to better crises response. ICT has enhanced disaster management efforts within organizations in the form of Knowledge Management System (KMS). KMS has seven major categories which are: Expert Systems,

Table 2. Major disasters that hit Malaysia

Year	Disaster	Killed	Injured	Total affected	Damage (m)
2008	Flood			10,210	
2008	Landslide	11	15	1422	
2007	Flood	33		158,000	225
2006	Flood	19		138,000	343
2005	Flood	17		100,000	66
2005	Mud flood	3		2793	
2004	Flood	13	0	15,000	
2005	Flash dlood	4	0	600	
2004	Tsunami	80	767	5063	14.6
2005	Wild fires	0	0	0	–
2004	Storm	0	0	1000	–
2004	Storm	1	0	40,000	–

Source www.adrc.asia.com

artificial intelligence and knowledge based management system (KBMS), Groupware (Computer Supported Collaborative Work), Document management systems, Decision support systems (DSS). Semantic networks • Relational and Object oriented databases, Simulation tools.

As part of legislation, the Malaysian government has set guidelines for disaster response and the responsibilities of various agencies nationally and internationally [15]. Extra attention was given after learning from shortfalls in disaster management, which gave awareness and spurred training initiatives that suited contemporary practice [16]. Research shed light on the role of developing knowledge and cultural values to resolve crises. It can be concluded that restorative community,[1] artificial community,[2] unselfish community,[3] emergence[4] in case of disaster, and emergence behavior[5] to operate in an environment are all part of effective disaster management [16]. Furthermore, a report published suggests that stakeholder participation is essential in the occurrence of disasters; a culture of prevention and reconstruction should be fostered in communities because natural hazards and disasters might happen from several angles, and people should be prepared for the threats [17]. It is essential to increase and promote community resilience in the case of disaster; therefore, holistic planning is needed to minimize people's vulnerability. Many have not considered the effectiveness of a culture of preparedness in reducing disaster impact [18].

[1] Restorative community: a collaborative effort with a mission to build the capacity and sustainability of organisations, initiatives and networks.

[2] Artificial community: accidentally come together for short time.

[3] Unselfish community: deliberate pursuit of the interests or welfare of others or the public interest.

[4] Emergence: the act of emerging a disaster response structure.

[5] Emergent behaviour: communities operate in an environment, forming more complex behaviours as collective.

Only after the collapse of the highland towers in December 11, 1993, did the Malaysian government realize the weakness of disaster management in the country [16]. A lack of responsibility and local expertise increased the tragedy of this landslide, which revealed insufficient standardized rules and specialized rescue operations in Malaysia. Since the tragedy of the condominiums, the Malaysian government has been working to develop guidelines on the international and national level to promote better disaster relief management among agencies and communities [16]. Policies and mechanisms for national disaster and relief management have been formed. Malaysia's Disaster Assistance and Rescue Team (SMART) is responsible for natural and technological disasters as Directive 20 of their policy purposely states [19]. They are to perform the following functions:

- Mitigate the effects of various hazards.
- Prepare for measures that will preserve life and minimize damage to the environment.
- Respond during emergencies and provide assistance.
- Establish a recovery system to ensure the affected community's return to normalcy.

Directive 20 is handled by the Standard Operating Procedure (SOP) that deals with natural disasters that might occur suddenly, such as landslide and flood, or technological disasters, like factory explosion and fire. There are three levels of incidents [16]:

- Level 1 Disaster: local incidents within control.
- Level 2 Disaster: more serious incidents covering a wide area.
- Level 3 Disaster: any incident that is more complex and affects wider areas (more than two provinces).

However, Malaysia is still in the stage of restructuring and reorganizing their natural disaster management mechanism, it only adopted the Hyogo Framework of Action HFA in November 2005 during a committee meeting that resulted from the ASEAN Agreement on Disaster Management and Emergency Response (AADMER-HFA input). The committee meeting chaired by the Deputy Prime Minister. According to the statement by HFA; a country must consider acceding to ratifying the relevant international legal instrument (e.g. the humanitarian Charter and Minimum Standards, international human rights law, refugee law and the Code of Conduct for the International Red Cross and Red Crescent Movement and Non-Governmental Organisations) [2].

4 Preparedness and Prevention

Disaster preparedness is a concern for the Malaysian government and citizens. As the country develops, this concern continues to evolve among various stakeholders regarding the risks of disasters and community readiness [13]. A report published explains the types of disasters that are likely to occur based on location with an aim to increase awareness and preparedness among the community; a culture of prevention is an effective approach for a safer community [6].

The United Nations Development Programme (UNDP) and Asia-Pacific Development Information Programme (APPDI) in 2007 described the stages of a disaster management cycle. The pre-disaster stage is an essential component of any framework and refers to any disaster handled holistically rather than singularly. The during-disaster stage covers implementation of safety measures and early warning systems. The response phase is the implementation of the action plans [6]. Details of the stages are laid out below in Table 3.

Another research [13] further examines the role of culture in disaster prevention and preparedness among the Malaysian community. It states that lack of funds has caused a crisis inadequate training for disaster preparedness. Therefore, these researchers have proposed solutions to modulate and assess the effectiveness of training programs. However, the question remains whether Malaysians have enough knowledge to respond to any kind of disaster that might occur and to survive during disasters. Along the same line, research based on a survey of 1000 questionnaires found that although people claim to be prepared for a disaster, vulnerability still exists among the majority of Malaysians. The questions were divided into three sections: General information about respondents, Experience and Expectations on Disasters, Perceived Disaster Preparedness. Majority of the respondents did not give an appropriate level of concern. This reflects that they are unsure of risk posed by geo-hazard phenomena [13].

A study by [8] concludes several points about ways to reduce the number of deaths during a disaster, relying on lessons from the experience of Indonesia and Sri Lanka during the 2004 tsunami. Improving public awareness has to be taken into consideration in the future; an early warning could have saved thousands of lives. Additionally, environmental factors such as modern progress, ruthless destruction of natural defenses such as coral forests and mangrove swamps, and building oceanfront hotels and villas in violation of coastal conservation legislation; were not taken into consideration could have saved many lives. Social preparedness for such a disaster was absent.

Table 3. The stages of a disaster management cycle

Pre-Disaster stage	During-Disaster stage	Post-Disaster stage
Mitigation: any activity that reduces either the chance of a hazard taking place or a hazard turning into a disaster **Risk reduction**: anticipatory measures and actions that seek to avoid future risks as a result of a disaster **Prevention**: avoiding a disaster **Preparedness**: plans and preparations made to save lives or property and response and rescue service operations	**Response**: includes actions taken to save lives, prevent property damage, and preserve the environment during emergencies or disasters	**Recovery**: includes actions that assist a community in returning to a sense of normalcy after a disaster

Source Asia-Pacific Development Information Programme (APPDI)

Furthermore, it is obvious from the 2004 tsunami that adequate funding does not guarantee better disaster management. Local communities and governments should take responsibility to engage and arrange proper education and planning for such incidents.

5 A Culture of Survival

Disaster Risk Reduction (DRR) is a set of implementation measures that have been used to decrease the risk of human and material loss that could otherwise be devastating for a community in a natural disaster. Several disasters over recent years such as the Indian Ocean Tsunami (2004), Hurricane Katrina (2005), Italy's earthquake (2009), and Pakistan's floods (2010) have shown an increase in disasters resulting in the damage of many lives, properties, economies, and facilities. The noticeable increase of natural hazards and disasters also reveals an increase in human vulnerability [20]. The vulnerability of communities in natural disasters depends on physical and social factors [21]. Anthropologists argue that factors like religious beliefs, social values, traditions, and attachment to a place are priorities to people during hazardous situations and affect outcomes [20]. For example, Indigenous communities survived the 2004 tsunami who had been familiar with the location and area; whereas, migrants and tourists did not fare as well [20]. The Moken community in Thailand identified the signs of unusual bird activity to indicate the arrival of the tsunami and moved away from the sea to a safer area [22]. However, Sri Lanka is an example of a lack of knowledge of tsunamis. The people moved toward the sea rather than moving away as they saw the low tide indicating an impending tsunami. Evacuating the danger zone was unfamiliar to their community.

Furthermore, the role of culture in risk of disaster was further explained [23]. There is a strong link between livelihood and culture with a number of factors that impact people's vulnerability, such as access to resources, strategies of survival, and control of resources. Also, gender roles and traditional politics influence group behavior and disaster management. It was suggested that sometimes unsustainable and unproductive livelihood patterns continue because of traditional and societal habits that are passed on to future generations.

To illustrate the earlier claims, another case of disaster management worth mentioning is from Simeulue, a small island of Indonesia, which is located about 100 km from the epicenter of the 2004 tsunami. Although about 5500 houses were destroyed in the tsunami, only seven deaths were reported. This apparently happened because seagrass, coral reefs, and mangrove forests softened the waves and also because the cultural habit of the local people, transferred from generation to generation, was to recognize indications of an impending tsunami and run to warn others to take the highest hill. The rest would follow and warn others consequently. Wetlands International-Indonesia Program 2005 proved this procedure to be very effective in a tsunami; however, such a simple technique has never been used or might have been forgotten by the people of the area of Aceh and North Sumatra [8].

The ways of dealing with disasters by using everyday products to promote knowledge about how to handle natural disasters were discussed [24]. It was explained how products like movies, T-shirts, and stories like Noah and the ancient flood could be part of disaster management. It is mass marketing for people's benefit. The researcher believes that sociologists and disaster researchers can brainstorm unique ways to increase dialogue on the role of culture in disaster management.

6 Conclusions

By focusing on Malaysia as one of the Asian-Pacific countries that suffer from frequent hazards, the role of culture in disaster management becomes clear. A review of case studies from other regions was used in this discussion to compare the roles of key players in the occurrence of natural disasters and to show different communities' reactions to incidents. The research concludes that the Malaysian government has been working to improve and structure a culture of disaster management to minimize the number of deaths and property loss. Relying on examples from different natural disasters in recent years, it has been realized that government action alone is not enough to reduce loss and destruction; community and culture factors into disaster preparedness and management. Statistics show that the Malaysian community has a lack of knowledge in response to natural disasters, and knowledge has proven to be an effective factor of survival in other studies. It is essential to educate people, especially those who live in areas where future tragedies might happen, to be prepared and knowledgeable, promoting a culture of prevention and survival. It is recommended for future studies to share an awareness of disaster management among Malaysians to create a responsive community prepared for any future hazard and disaster. The culture of prevention and mitigation should be implanted in the community by suggesting further studies on Malaysian's behaviour, relying on the examples of previous tragedies and other communities.

References

1. Dorasamy, M., Raman, M., Muthaiyah, S., Kaliannan, M.: Knowledge-based emergency management information systems : mapping DERMIS principles. In: Knowledge Management International Conference (KMICe), pp. 237–242 (2012, no. July)
2. Mat Said, A., Ahmadun, F.-R.: Disaster Management in Malaysia: evolution, Development and Future Challenges (2007)
3. IFRCRCS.: Disaster management and risk reduction: strategy and coordination. In: International Federation of Red Cross and Red Crescent Societies, pp. 1–8 (2010)
4. ADRC.: Natural Disaster Data Book 2009 (An Analytical Overview) (2011)
5. Rodriguez, J., Vos, F., Below, R., Guha-Sapir, D.: Anual Disaster Statistical Review 2008—the numbers and trends (2009)
6. Wattegama, C.: ICT for Disaster Management. United Nations Development Programme—Asia-Pacific Development Information Programme (UNDP-APDIP) and Asian and Pacific Training Centre for Information and Communication Technology for Development (APCICT)—2007, p. 48 (2007)
7. Isdr.: Developing early warning systems : a checklist. In: Third International Conference on Early Warning, no. March, pp. 1–13, (2006)

8. Athukorala, P., Resosudarmo, B.: The Indian ocean tsunami : economic impact, disaster management and lessons. Asian Econ. Pap. **4**(1), 1–39 (2005)
9. Economist.: Relief but little rebuilding|The Economist. Asia's tsunami: Relief but little rebuilding, 2005. [Online]. Available: http://www.economist.com/node/5327849. Accessed: 19 Jul 2017
10. Deutsche Presse-Agenture.: Governments not interested in warning system (2005)
11. Thanawood, C., Yongchalermchai, C., Densrisereekul, O.: Effects of the December 2004 tsunami and disaster management In Southern Thailand Chanchai Thanawood, Chao Yongchalermchai and Omthip Densrisereekul. Sci. Tsunami Hazards **24**(3), 206–217 (2004)
12. UN.: The millennium development goals report (2006)
13. Magiswary, D., Murali, R., Saravanan, M., Maniam, K.: ICT and disaster preparedness in Malaysia: An exploratory study. WSEAS Trans. Inf. Sci. Appl. **7**(5), 735–748 (2010)
14. Hassan, N.A., Hayiyusuh, N.-A., Nouri, R.: The implementation of knowledge management system (KMS) for the support of Humanitarian assistance/disaster relief (HA/DR) in Malaysia. Int. J. Humanit. Soc. Sci. **1**(4), 103–112 (2011)
15. Shaluf, I.M., Ahamadun, F.-R.: Major hazard control: the Malaysian experience. Disaster Prev. Manag. **12**(5), 420–427 (2003)
16. Roosli, R., O'Keefe, P., Mydin, A.O.: Evolution of disaster planning and housing in Malaysia: a reviews. World Appl. Sci. J. **21**(7), 945–959 (2013)
17. Quarantelli, E.L.: Organizational behavior in disasters and implications for disaster planning (1986)
18. Drabek, T.E.: Revisiting the disaster encyclopedia. Int. J. Mass Emerg. Disasters **17**(2), 237–257 (1999)
19. Aini, M.S., Fakhru'l-Razi, A., Daud, M.: Evolution of emergency management in Malaysia. J. Contingencies Cris. Manag. **9**(1), 46–53 (2001, March)
20. Kulatunga, U.: Impact of culture towards disaster risk reduction. Int. J. Strateg. Prop. Manag. **14**(4), 304–313 (2010)
21. Manandhar, R., McEntire, D.A.: Disaster and development. Dev. Pract. **20**(7), 904–905 (2010)
22. UNISDR: Indigenous knowledge for disaster risk reduction. Indig. Knowl. Disaster Risk Reduct. **97**(1) 12–21 (2008)
23. Cahn, P.S.: Náyari history, politics, and violence: from flowers to ash. Am. Ethnol. **29**(4), 1017–1018 (2002)
24. Wachtendorf, T.: Exploring The Popular Culture Of Disaster. Chicago Illinois (1999)

Numerical Investigation of Out of Plane Behavior of Hollow Core Inverted Waffle Slab System

Nazaib Ur Rehman(✉), Bilal Yousaf, and Syed Ubaid Ullah Shah

Department of Civil Engineering, University of Engineering and Technology, Peshawar, Pakistan
{nazaibkhan,ubaidkhan779}@yahoo.com, engr.bilal.yusuf@gmail.com

Abstract. Long span reinforced concrete slabs results in heavy structures and therefore technologies like waffle slabs are required to provide slabs with smaller weight to span ratios. But in waffle slabs the complications in placing of reinforcement as well as in installation of formwork leads to costly and difficult to in construct slab systems. A new slab system, that is basically Inverted waffle slab system with hollow core inside, can avoid this problem by placing permanent "waffle pods" in the empty spaces of the joist. Waffle pods should have light weight, reasonable strength, fire resistance and economical. Gypsum as one of the many possible solutions was used as a waffle pod to fill these spaces. In this paper a comparative study of behavior of hollow core inverted waffle slab (IWS) having RCC on top, conventional waffle slab (CWS) and conventional solid slab (CS) has been carried out analytically. For this purpose these slabs were modeled and analyzed using finite element application SAFE V.12, and the output calculations were compared. In this study the dimension of the slabs panel were chosen as 30 × 30 ft. and were modeled with same construction material and loading conditions. Various parameters were studied using these models such as maximum deflection, maximum bending moment, and maximum area of steel. This research shows that hollow core inverted waffle slab system may overcome the problem of formwork for conventional waffle slab system.

Keywords: Waffle core · Slab · Finite element modeling · Cost analysis Reinforced concrete

1 Introduction

Slab is a horizontal structural component with top and bottom surfaces parallel or near so, the depth of this structural element is very small compared to that of its span.

There are four major types of slab system, **(a)** Slab beam system, it is the most economical system but it is not appealing to the people due to the beams hanging from the slabs, requires more depth and hence there will be increase in storey height, **(b)** flat plate and **(c)** flat slab both are appealing but they are uneconomical and less earthquake resistant, by which they are not preferable in high seismic zone areas, the last system

M. Awang and M. H. Isa (eds.), *The Advances in Civil Engineering Materials*,
Lecture Notes in Civil Engineering 19, https://doi.org/10.1007/978-981-13-2511-3_8

that is **(d)** waffle slab system is both appealing and economical as compared to other slab systems but it has the disadvantage of complicated formwork. For all these suspended slabs, there are a number of designs to improve the strength to weight ratio. In all cases the top surface remains flat, and the underside is modified. According to P. F. Schwetz, the behavior of reinforced concrete waffle slabs still raises some questions due to their complexity which has led some researchers to undertake numerical and experimental studies (Abdul-Wahab and Khalil 2000; Schwetz 2005, 2013; Selistre 2000; Soares 2003, Ajdukiewicz and Kliszczewicz 1986).

According to P. F. Schwetz, Highly developed and astute structural solutions are going on an increasing demand in the daily work of structure designers due to the evolution of architectural designs and new construction management concepts. Therefore, it is requisite to evaluate a variety of structural systems, seeking to find the solution that can best deliver economic viability, speed and versatility of application.

According to Dias R.H, the utmost part of concrete is consumed by slabs in multi-storied reinforced concrete structures, This utilization reaches up to 67% of the total volume of the structure in case of the solid slab. Therefore, it is impFor Conventional two way slab systemortant to study different types of slab to be used in multi-story buildings to find solutions that are both technically viable and less expensive.

The use of waffle slabs is proved to be an interesting option. This system can be defined as a reinforced concrete slab with equally spaced ribs parallel to the sides, having a waffle appearance from underside.

In fact, as the gradual development of numerical algorithms that allow the design of this type of structure is quite recent, many questions still remain with regards to the quantification of the slab stresses and their adequate modeling by the design methods and the mathematical models currently used.

Conventional structural arrangements made of solid slabs and concrete frames remain the preferred model for reinforced concrete structures due to the ease of its formwork and its construction methods, but the main issue of this system is the short span and also not aesthetically appealing not preferred by most architects, so a number of research work has been done numerically and experimentally on analysis and design of waffle slab system, various procedures and standards has been developed to guide the use of waffle slabs construction system due to the increase demand of waffle slab system.

So demand of waffle slab system is increasing to that of conventional slab systems, due to its structural stability and architectural view, as some of the studies by Akshay Kumar & Shilpa and Anjaly Somasekhar & Preetha Prabhakara shows the superiority of waffle slab over solid slab which can be inferred from their study results, that in waffle slabs bending moments, area of steel and deflection reduces up-to 22, 36 and 67% respectively, also studied by Kennedy and Nabil F. Grace shows that not only the waffle slab has advantages over the conventional solid flat plate, but also the pre-stressed waffle type bridge is found to be much more efficient in carrying load as compare to that of pre-stressed bridge with uniform thickness slab, they studied the effect of orientation of rib in the load carrying capacity of waffled slab. Their results indicated that the orthogonal shaped waffle slab has a superior ultimate load carrying capacity of 20% higher than the non-orthogonal (45°) waffle slab. But the complication in formwork in waffle slab system create problems on the site. Temporary Waffle pods are being used in waffle slab system to fill the gaps between the joists which makes it difficult for labors and becomes uneconomical due to high expense of formwork.

A new slab system known as hollow core Inverted waffle slab system may overcome this formwork complication by placing permanent Waffle pods, in the vacant spaces in between the joists. Now the pods to secure the empty spaces of slab system, must have light weight, reasonable strength, fire resistance and economical. Gypsum as one of the many possible solutions was used as a waffle pod to fill these spaces. This work verifies through numerical analysis (Finite Element Method) of hollow core inverted waffle slab stiffened with ribs, and comparative study of three slabs has been carried out.

2 Finite Element Model

The modeling, analysis and design of all the three distinct slab system has been done by using finite element software SAFE V.12. Slab ID's, Slab types and the geometric details of three different slabs are given in Table 1.

For Conventional two way slab system

According to ACI 318-14 **(L/B < 2)**
h_{min} = Perimeter/180 = $2(l_a + l_b)/180$
So $h_{min} = 30 \times 4/180 = 0.67$ f = 8 in so minimum slab depth is 8 in.

Table 1. Dimensions of three different Slab systems

Slab ID	Slab type	Slab area (ft^2)	Slab depth (in)	Beam/Rib length (ft.)	Beam/Rib width (in)	Beam/Rib depth (in)	Column dimensions (in)
CS	Conventional solid slab	30 × 30	8	30	12	18	18 × 18
CWS	Conventional waffle slab	30 × 30	3	30	6	12	18 × 18
IWS	Hollow core inverted waffle slab having PCC on top	30 × 30	3	30	6	12	18 × 18

3 Loadings

Only gravity loads are considered and the effect of lateral loads has not taken into account. The loadings for three types of slab, are live load that of residential building taken as 40 psf and the dead load due to floor finishing as 30 psf and unit weight of reinforced concrete has been taken 150 lb/ft^3 for calculating self-weight of slab. Load of PCC on top of inverted waffle slab and load of gypsum block has been taken as 37.5 and 34 psf for Hollow core inverted waffle slab system only.

3.1 Loading Combination

Two loading combination has been used during analysis of model, namely Factored load and Service load.

(a) **Factored Load**
Factored load was used as according to **ACI 318-14** Table 5.3.1—**Load combinations** which is 1.2D.L + 1.6L.L and was used in calculating maximum bending moment and maximum area of steel.
(b) **Service Load**
Service load has been used for calculating deflection only.

4 Material Properties

Two different construction materials, Concrete and Gypsum were used and steel was also used as reinforcing material and each material properties are given in Table 2.

Table 2. Material properties

Material	Property	Value
Concrete	Compressive strength	3 ksi
	Young modulus	3200 ksi
	Poisson ratio	0.2
	Shear modulus	1300 ksi
	Unit weight	150 psf
Reinforced steel	Tensile strength	60 ksi
Gypsum	Young modulus	29,000 ksi
	Unit weight	49 Psf

5 Output Result

The deflection values were taken at the mid-point of slab, Maximum moment and Maximum area of steel values were taken from both the middle and column strip. The numerical value of each Parameter (deflection, maximum moment and maximum area of steel) for three different types of slabs and percent deviation in different parameter between CS & IWS and CWS & IWS are shown in Table 3.

(a) **Graphical representation of output result**
The deflection, maximum moment and maximum area of steel parameters values are plotted in Figs. 1, 2 and 3 respectively.
(b) **Quantity Comparison**
The quantity of concrete and total weight of steel used in different type of slabs are shown in Table 4.

Table 3. Comparison of results in CS, CWS and IWS system

Parameters	Conventional Slab system CS	Waffle Slab system CWS	Hollow core inverted waffle slab system IWS	Percent deviation between CS and IWS	Percent deviation between CWS and IWS
Deflection (in)	1.03	0.59	0.88	14	−49
Maximum moment (kip-ft.) at middle strip	149.81	17.60	46.4	69	−164
Maximum moment (kip-ft.) at column strip	72.24	16.95	53.2	26	−214
Maximum area of steel (in^2/ft.) at middle strip	0.33	0.192	0.51	−54	−166
Maximum area of steel (in^2/ft.) at column strip	0.29	0.199	0.58	−100	−191

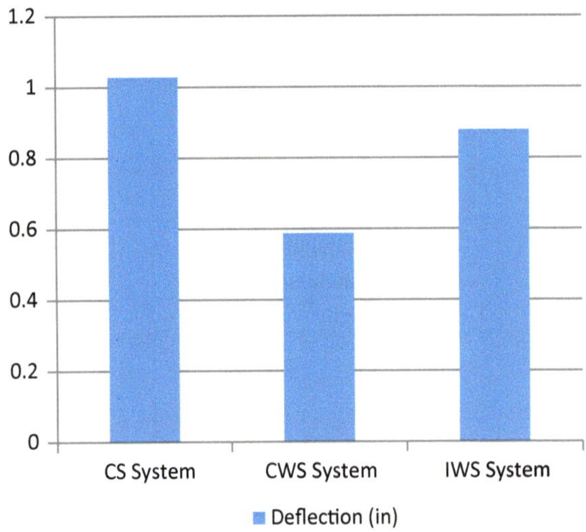

Fig. 1. Demonstrates deflection parameter for three distinct type of slabs

6 Result and Discussion

As long span reinforced concrete structures are not possible with CS system without pre-stressing, so only considering CWS and IWS system. Overall, an increase in each of parameter (deflection, maximum moment and maximum area of steel) has been shown by the IWS system to that of CWS system, having 49% in deflection, 164 and 214% in maximum moment demand at middle and column strip, 166 and 191% increase in maximum area of steel at middle and column strip. Results were also

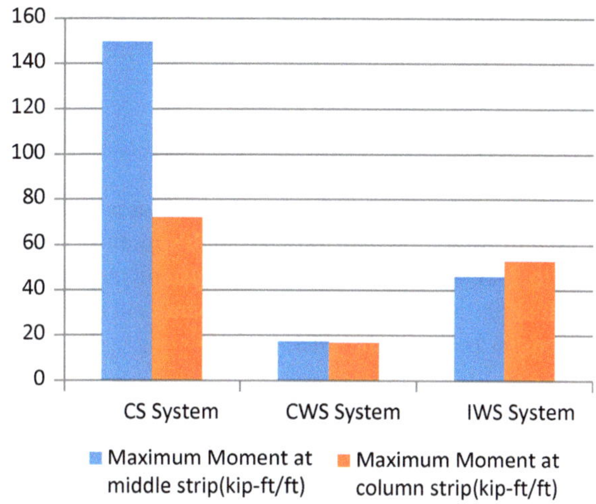

Fig. 2. Demonstrates maximum moment parameter for three distinct type of slabs

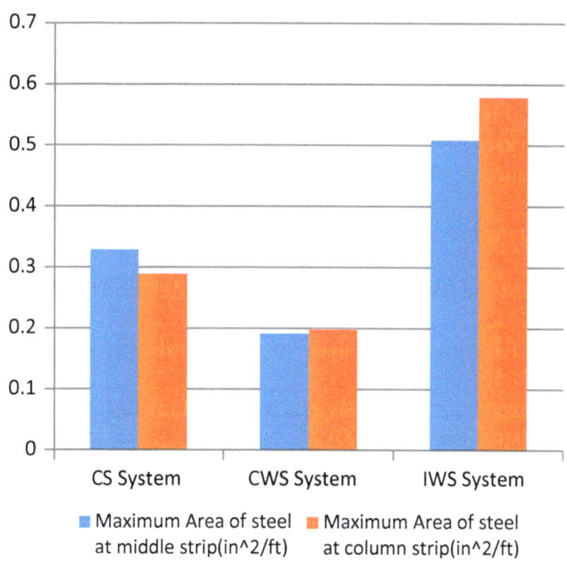

Fig. 3. Demonstrates maximum area of steel parameter for three distinct type of slabs

Table 4. Comparison of cost in CS, CWS and IWS system

Slab type	Concrete quantity (ft^3)	Total weight of steel (lbs.)
CS system	600	3610
CWS system	621	959
IWS system	846	3938

obtained for quantities, including 36% increase in concrete quantity and 310% increase in weight of steel of IWS system over that of CWS system.

7 Conclusions and Recommendation

It is concluded that long span reinforced concrete structures are not possible with CS system. So CWS system is the one way to make such structures, but cost and difficulty in formwork leads to another problem. So IWS system may avoid complication in formwork (only with slight increase in moment demand and reinforcement) by placing permanent waffle pods in between the joist.

Acknowledgements. The authors are thankful to the administration of Civil Engineering Department, University of Engineering and Technology, Peshawar for providing digital library access.

References

ACI Committee 318.: Building code requirements for structural concrete: (ACI 318–99); and commentary (ACI 318R-99). American Concrete Institute, Farmington Hills, Mich (1999)

Dias, R.H.: Numerical Analysis of Pavements of Buildings in Ribbed Slabs (2003)

Grace, N.F., Kennedy, J.B.: Dynamic characteristics of prestressed Waffle Slabs. J. Struct. Eng. **116**(1990), 1547–1564 (1990)

Kennedy, J.B.: Orientation of ribs in Waffle-Slab skew bridges. J. Struct. Eng. **109**(1983), 811–816 (1983)

Kumar, A., Pal, S.: Parametric study and analysis of waffle and solid slab. J. Civil Eng. Environ. Technol. **2** (2015)

Schwetz, PF.: Numerical and Experimental Study of a Waffle Slab Designed to serve as a Tennis Court Floor (2013)

Schwetz.: Theoretical-Experimental Analysis of a Reduced Model Ribbed Slab subjected to Linear Loading (2005)

Selistre.: Theoretical-experimental analysis of a reinforced micro concrete ribbed slab subjected to uniformly distributed loading (2000)

Soares.: Experimental Analysis of Ribbed Mushrooms from Concrete Armed with Metal Pillars (2003)

Somasekhar, A., Prabhakaran, P.: Analysis of reinforced concrete Waffle Slabs with openings. Int. J. Emer. Technol. Adv. Eng. **5** (2015)

Wahab-Abdul, H.M.S., Khalil, M.H.: Rigidity and strength of orthotropic reinforced concrete Waffle Slabs, J. Struct. Eng. **126**, 219–222 (2000)

Application of Phase Change Materials (PCMs) in Building Walls: A Review

Zeyad Amin Abdo Saeed Al-Absi(✉), Mohd Hafizal Mohd Isa, and Mazran Ismail

Architecture Department, School of Housing, Building and Planning, Universiti Sains Malaysia 11800, Penang, Malaysia
zeyadaminalabsi@yahoo.com, {hafizal,mazran}@usm.my

Abstract. The rapid growth of the population and overcrowded of urban areas forced building construction sector to focus more on economic consideration rather than climatic requirements. As a result, achieving comfortable living spaces was fully dependent on mechanical systems, which caused more energy consumption and greenhouse gas emissions. Passive design strategies become an attractive alternative to tackle this problem and to reduce the negative impacts on our planet. Phase Change Materials (PCMs) store superior amount of latent heat when changing their phase compared to sensible heat. PCMs application in buildings helps to lower indoor temperature and reduce temperature fluctuation and total hours of overheating. Additionally, PCMs can also absorb the internal heat dissipated by household equipment, lightings and occupants' activities. All this help to improve building indoor environment. This paper aims to review PCM applications for building walls. It was found that PCMs might be incorporated to building walls as pre-fabricated PCM-enhanced elements such as PCM-enhanced wallboards, panels, bricks and blocks. In addition, it might be incorporated on the site to fresh mixtures such as concrete, plaster and mortar and applied to buildings. The thermal performance of building walls was improved in all cases resulting in heating and cooling load reductions. Moreover, many researchers found that applying night ventilation further improves the thermal performance of PCMs in building walls.

Keywords: Phase change materials (PCM) · Building walls · Thermal energy storage (TES) · Passive building materials

1 Introduction

Buildings accounted for 40% of global energy consumption and the largest use of energy in buildings (i.e. 60%) is for heating and cooling. Hence, they offer a good opportunity to reduce energy consumption through the utilizing of other alternatives such as passive design strategies [1]. Thermal energy storage (TES) acts as a heat sink by storing energy for later use. This technology helps to increase the effective use of thermal energy equipment and systems and to improve heat exchange for energy efficiency in buildings, hence reduce the energy consumption. In addition, it helps to tackle the mismatch between the energy's supply and demand [2]. Type of TES is known as Phase Change Materials (PCMs) that is capable to utilize the latent heat

M. Awang and M. H. Isa (eds.), *The Advances in Civil Engineering Materials*,
Lecture Notes in Civil Engineering 19, https://doi.org/10.1007/978-981-13-2511-3_9

absorbed or released during the phase change of the materials. PCMs can control the ambient temperature within a specific range corresponds to the PCMs' phase transition temperature. If the phase transition temperature of the material matches the required comfort temperature, it can help to absorb the extra heat leading to more stable and comfortable indoor environment, hence reduce the required heating and cooling loads. PCMs are attractive for many researchers and have been successfully implemented in buildings for thermal management due to their wide range of categories, several melting/frizzing temperature, relative steady melting/frizzing temperature, and high thermal density with small temperature intervals and negligible volume change [3]. PCMs can be classified to organic PCMs, inorganic PCMs, and eutectic PCMs, which further divided based on the several components of the PCMs (Fig. 1).

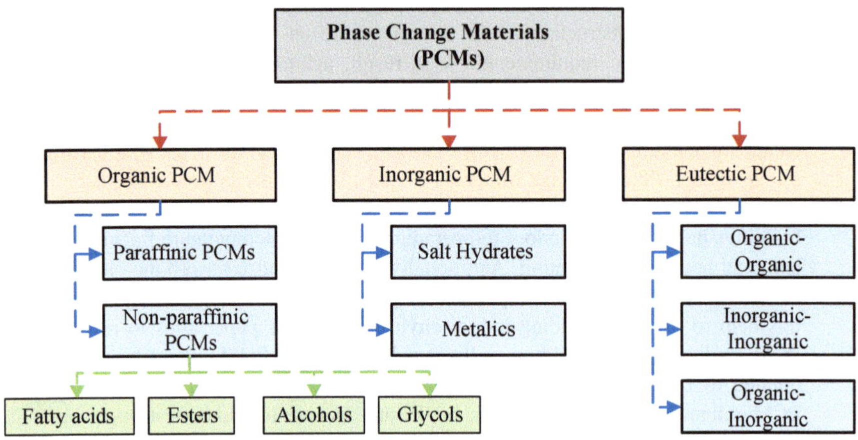

Fig. 1. PCMs classification

In addition, PCMs can be integrated into building materials by different strategies such as; direct incorporation, immersion, encapsulation and stabilization. [1, 4]. In buildings, walls and ceilings offer large areas in each zone for passive heat transfer [5] in which PCMs can be incorporated. Therefore, the objective of this paper is to review PCMs applications for building walls and the methods used for the incorporation of PCMs into buildings.

2 PCM Application in Buildings Walls

Based on the literature, PCMs might be applied to building walls using different methods including prefabricated elements or mixing on the site. These methods are discussed below.

2.1 PCM-Enhanced Wallboard and Panels

PCM wallboards and panels are of the most common ways used to integrate PCM into buildings due to their ease of installation. For example, liquid PCM was absorbed into gypsum board by soaking [6], microencapsulated PCMs were incorporated into wood-plastic composites to improve its thermal properties [7], two types of PCMs with different melting temperatures were macro-encapsulated by aluminum panels and tested for the summer and winter [8], PCM composite wall made by incorporation of shape-stabilized PCM particles in a polymer matrix [9] and Shape-stabilized PCM wallboard made of paraffin wax PCM and high-density polyethylene [10]. In all cases, the addition of PCM have shown improvement in thermal behaviour of building materials as well as the indoor thermal environment of buildings.

Zhu et al. [11] performed simulations to investigate the use of two different double-layer shape-stabilized phase change materials (SSPCMs) wallboard to save energy in five typical climate regions in China. The SSPCMs wallboards with the thickness of 30, 40 and 50 mm, were attached on the inner and outer surfaces of a common concrete wall in the south of the simulated room, while the other three walls, floor and ceiling were thermally isolated using insulation materials. The air-conditioner operated from 7:00 to 18:00 (i.e. office building) with set-point temperature of 26 °C for summer and 18 °C for winter in the regions that required heating or cooling. In this work, the external PCM layer was used to reduce the cooling load in summer or to reduce heat gain if there is no need for cooling. In contrast, the internal PCM layer was used to reduce heating load in winter or to reduce heat loss if there is no need for heating. In most cases, the maximum reduction is achieved with the highest thickness of SSPCM wallboards. This might suggest the largest PCM content in the PCM-wallboard panel. In addition, the optimal melting temperature of the external layer was mainly influenced by the outdoor air temperature while the internal layer was mainly influenced by the indoor air temperature. Furthermore, they stated that smaller average outdoor temperature had better effects on energy savings in summer, while bigger average outdoor temperature had better effects in winter.

Evola et al. [12] simulated a typical office building containing PCM wallboard in the partition walls. The PCM wallboard, with the thickness of 20 mm, is made of aluminium honeycomb matrix, which contains 60% of approximate 5 μm diameter microencapsulated paraffin, and covered with two aluminium sheaths. The partition walls are composed of 70 mm thick gypsum board to which the PCM wallboards are fixed. The results show that PCM wallboard reduced the peak operative temperature by about 1 °C. In addition, the surface temperature and the daily temperature swing for the east wall were reduced by 1.7 – 2.8 °C respectively and the time for peak surface temperature was shifted by 1.5 to 2 h.

Kong et al. [13] investigated the cold storage performance of PCMs in the summer for the middle latitude region of China. Aluminium sheets, Fig. 2 were used to produce macro-encapsulated PCM panels containing two types of PCM (i.e. capric acid used for the external wall surface and a mixture of capric acid and 1-dodecanol used for the internal wall surface to reduce the super-cooling effect of the capric acid and allow for frizzing within small range in the internal environment). Internal fins were used in the aluminium panels to increase the heat transfer into the PCM and to support the panels' structure. Three rooms (2000 × 2000 × 2400 mm) were constructed from perforated bricks for the walls and insulation sandwich board for the roofs. PCM panels were installed in the outer surfaces of one room (PCMOW) and the inner surfaces of the second room (PCMIW), while the third room was the base case. A door and a window were built in each room to allow night ventilation. They concluded that PCM panels helped to reduce the temperature by more than 1 and 2 °C for PCMOW and PCMIW respectively and to delay the time for peak temperature by 2–3 h. In addition, they stated that the use of PCM in the outside surface has more effect of thermal insulation than temperature regulator.

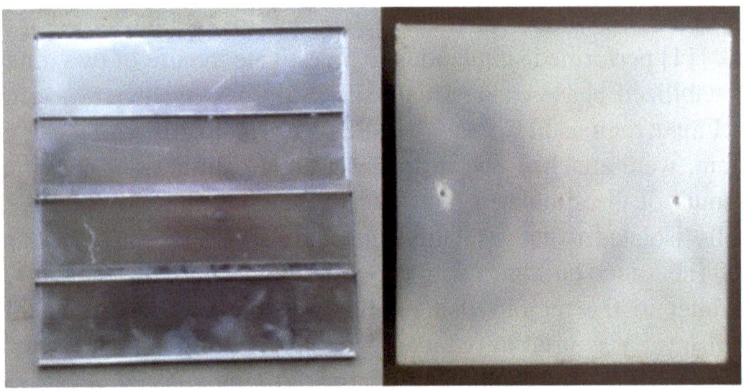

Fig. 2. PCM aluminum panels [13]

2.2 PCM-Enhanced Bricks and Blocks

PCM-enhanced bricks and blocks are subjected to investigate by many researchers. They offer large volume to incorporate PCMs. Incorporation of PCMs might be done by immersion, macro encapsulation or shape-stabilization. For example, macro-encapsulated PCMs-bricks were produced by filling the PCM into the cavity of the bricks [14–17], PCM macro-capsules made of steel were incorporated into brick masonry wall [18] and shape-stabilized PCMs were incorporated into cement mortar to produce PCMs-bricks [19–21].

In an experiment, Vicente and Silva [22] investigated the effect of PCM macro-capsules incorporated into brick masonry wall and the addition of Extruded polystyrene (XPS) insulation to the PCM-brick masonry wall. A metal steel macro-capsule (i.e. 300 × 170 × 28 mm with 0.75 mm thickness) were filled with paraffin and inserted in fired clay bricks with horizontal hollow (Fig. 3). Two brick masonry walls were constructed with PCM macro-capsules and one with normal brick as a base case. An insulation layer, XPS 10 mm thick, was installed on the outer surface of one PCM-brick masonry wall. The three walls were tested between two climatic chambers, for imposed conditions and free float mode.

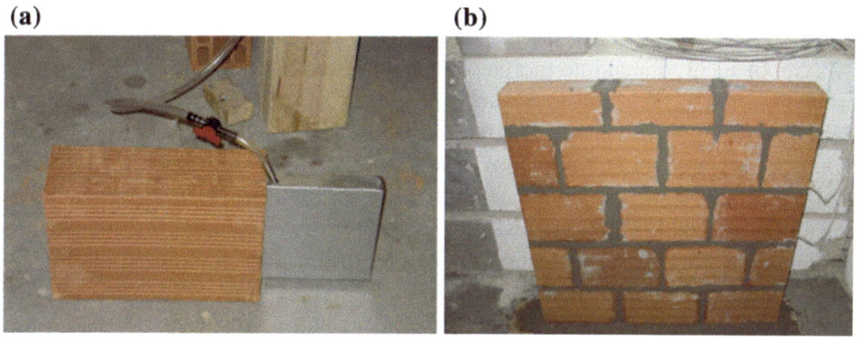

Fig. 3. **a** PCM macro-capsules; **b** wall specimen [22]

The results showed a delay of 3 h for the maximum peak temperature achieved by both PCM wall and insulated PCM wall compared to 1 h for the base case, which indicates how PCM can improve the thermal inertia of the wall. In addition, the PCM wall and insulated PCM wall reduced the temperature fluctuation by 5 and 8 °C (i.e. 50 and 80%) respectively. The insulated PCM-wall achieved the lowest maximum temperature and the highest minimum temperature; even though, the PCM wall resulted with superior improvements compared to the conventional wall. However, a comparison with insulated base case wall should be done to find out the role of the PCM with the insulation materials.

Wang et al. [23] investigated PCM bricks produced by incorporating shape-stabilized PCM into cement mortar (i.e. mass ratio was 37.5, 22.5 and 40% for portland cement, sand and shape-stabilized PCM respectively), which poured into 240 × 120 × 90 mm homemade wooden molds (Fig. 4). The PCM-bricks were used in one side of a ful-scale room of 3250 × 3860 × 2910 mm and compared to wall built with perforated vitrified bricks. The test were performed for the year-round. The results showed a reduction of 24.32% of the cooling load during summer and 10–30% of heating load during winter. In addition, when the air-conditioner was off, the heat lost from the interior wall surface can be reduced by 9–72%. They summarized that PCM-wall achieved better thermal behavior all the year round.

Fig. 4. **a** SSPCM; **b** SSPCM-bricks; **c** perforated vitrified bricks [23]

2.3 PCM-Enhanced Plaster

PCMs can be incorporated into mortars and applied to buildings walls as a PCM-plaster coating or casted to PCM-plaster wallboards. For example, PCM was incorporated into aerial lime and gypsum based mortars and tested as a coating in laboratory-scale prototypes [24], a novel PCM composite was integrated into ordinary cement mortar as partial replacement for fine aggregate and evaluated the thermal behaviour as a plaster in self-designed box [25], PCM-enhanced lime plasters were produced and evaluated to be compatible for renovation and retrofitting of existing traditional buildings in southern European climatic conditions [26] and a hybrid-PCM-plaster was developed by incorporating three types of PCM into cement mortar and tested as a coating in prototypes placed in a controlled climatic chamber [27]. This method makes it easy to integrate PCM into new buildings as well as to renovate existing buildings.

Lachheb et al. [28] investigated the thermal behaviour of new PCM-plaster composite as a component of passive solar walls. A sample with 10% microencapsulated paraffin was prepared and experimentally tested using guarded hot plates method for its thermo-physical properties. The results then used to validate the numerical investigation based on the enthalpy method using finite volume method. Both experimental and numerical results exhibit a similar behaviour, which indicated a good agreement. Then the numerical model was further used to investigate the effect of incorporation of PCM-plaster wallboard in building walls. The use of 10% PCM fraction within the plaster wallboard resulted in reduction of 2.7 °C and 11 Wm^{-2} in the wall inner surface temperature and heat flux respectively. Furthermore, by increasing the fraction of PCM to 15, 20 and 30%, they found that 20% of PCM greatly decreased the wall inner surface temperature to 25.8 °C and the heat flux to 16 Wm^{-2} compared to 28 °C and 35 Wm^{-2} respectively for normal plaster, while the 30% PCM achieved just a small reduction further to the 20% PCM. In addition, increasing the PCM-plaster wallboard thickness from 10 to 30 mm resulted in further reduction of 1.2 °C.

Sarı et al. [29] developed a novel cement based-composite phase change material (CB-CPCM) and tested its thermal performance as a plaster. The PCM was absorbed by cement using vacuum embedding method to prepare a form-stable CB-CPCM with maximum mass fraction of 28% without leakage. Thermal performance of the CB-CPCM plaster was studied using two polystyrene cubes (250 × 200 × 140 mm) coated internally with 5 mm-thickness of CB-CPCM plaster in one cube and ordinary cement in the other as a control cube (Fig. 5). The cubes were subjected to heat source until the indoor temperature raised 5 °C above the melting temperature then cooled until it was decreased bellow the frizzing temperature. Results showed the inner surface temperature of the cube with PCM plaster was lower and increased from 14.8 to 24 °C within 160 min compared to 125 min for the control cube. In addition, the time with comfortable temperature range (19–24 °C) was 100 min with PCM compared to 80 min without PCM. On the other hand, during the cooling process, the time required for the inner surface temperature to decrease from 24 to 19 °C was 90 and 65 min for the PCM cube and control cube respectively, which linked to the latent heat discharge. Similarly, the indoor temperature had a delay of 15 min with the PCM plaster to increase from 15 to 24 °C with an average temperature difference of 0.70 °C and a delay of 10 min to decrease from 24 to 19 °C with average temperature difference of 0.58 °C.

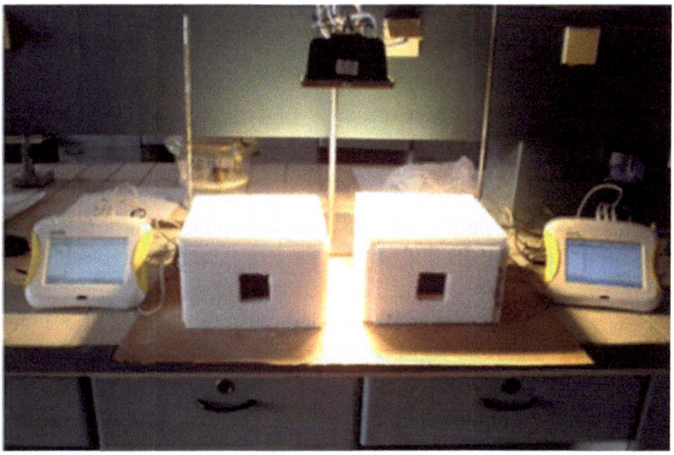

Fig. 5. Polystyrene cubes for thermal performance test [29]

3 PCMs and Ventilation

PCMs' performance and effectiveness are highly dependent on the surrounding environment. In some cases, the temperature at night may stay higher or just below the solidification temperature of the PCM for short period of time. Therefore, the PCM will not be able to fully transform from liquid to solid and release all the stored heat, which reduces its efficiency for the next day. Combining ventilation with PCM help to increase the rate of heat released by the PCM during night time, especially when the ambient temperature is close to the transition temperature of the PCM. Zhang et al. [30] found that ventilation can improve the thermal performance and energy saving efficiency of building utilizing PCM, though it is affected by ventilation periods. Furthermore, night ventilation is more effective in improving indoor thermal environment and reducing energy consumption for cooling. The optimum period for effective ventilation was found to be between 00:00 and 09:00. In addition, the more ventilation quantity used, the better performance of PCM can be achieved. However, the improvement in PCM performance becomes insignificant when more than 2 ACH was used. On the other hand, Ramakrishnan et al. [31] found that applying natural night ventilation with 8 ACH rate further decreased the peak indoor operative temperature, which indicates the role of night natural ventilation in discharging the stored heat in the PCM and preparing it for the next day cycle. The discomfort index (DI) illustrated that the sever discomfort hours (i.e. DI > 28 °C) for the five days were 36 h without the use of PCM and decreased to 28 h with the use of PCM. However, the sever discomfort hours further reduced to 25 h when applying natural night ventilation. The authors concluded that the combination of the PCM (optimum transition temperature and thickness of 29 °C and 30 mm respectively) and natural night ventilation resulted in reduction of 65% in the sever discomfort period during the heat wave. Furthermore, Barzin et al. [2] achieved 73% of weekly energy saving by applying cool night ventilation to charge the PCM rather than using air conditioning.

4 Conclusion

PCMs are suitable to be applied in building walls using different methods, which have shown improvement to thermal performance of building's walls, delay in peak temperatures and reduction in heating and cooling load. These methods are summarized below.

- PCMs wallboards and panels are one of the most common methods of PCM application on building walls due to their ease of installation in ether a new building construction or a retrofitting and renovation of existing buildings. They can be produced in mass quantities in controlled environment at the factories, which reduces the waste materials and provide higher quality control.
- PCMs-plaster also can be applied in new buildings construction and existing buildings retrofitting and renovation. However, this method required additional work at the site to be prepared, which may require some expertise or may not provide the quality of controlled environment production.
- PCMs-bricks and blocks can be produced at a controlled environment and transported to the site ready for construction. They can provide larger volume to incorporate PCMs, which overcome the limited thickness of wallboards, panels and plaster that may affect the amount of incorporated PCMs resulting in less thermal energy storage. However, their use is limited for new buildings construction or a major renovation of building's walls.

Additionally, night ventilation has shown great influence in the efficiency of PCMs products by increasing the rate of heat discharge, especially when ambient temperature is close to PCMs transition temperature. Therefore, effective night ventilation must be considered with any passive application of PCMs. To conclude, PCMs are promising as a passive technology for buildings sector. This study has reviewed methods of PCMs application into building walls. Each method was investigated separately in the literature. However, the effects of using multiple methods to introduce PCMs into building walls still unknown. The most common wall system is consists of a core (e.g. masonry and concrete walls) and two layers in both sides (e.g. wallboards and plaster). Incorporating PCMs to the whole wall system may perform better performance. The more PCMs amount is added, the more heat storage we can get. Therefore, further research in this point is required.

Acknowledgements. The authors would like to thank the School of Housing, Building and Planning, Universiti Sains Malaysia and Research University Grant (1001/PPBGN/814285) for the financial support for this research.

References

1. Akeiber, H., Nejat, P., Majid, M.Z.A., et al.: A review on phase change material (PCM) for sustainable passive cooling in building envelopes. Renew. Sustain. Energy Rev. **60**, 1470–1497 (2016). https://doi.org/10.1016/j.rser.2016.03.036

2. Barzin, R., Chen, J.J.J., Young, B.R., Farid, M.M.: Application of PCM energy storage in combination with night ventilation for space cooling. Appl. Energy **158**, 412–421 (2015). https://doi.org/10.1016/j.apenergy.2015.08.088
3. Iten, M., Liu, S., Shukla, A.: A review on the air-PCM-TES application for free cooling and heating in the buildings. Renew. Sustain. Energy Rev. **61**, 175–186 (2016). https://doi.org/10.1016/j.rser.2016.03.007
4. Wahid, M.A., Hosseini, S.E., Hussen, H.M., et al.: An overview of phase change materials for construction architecture thermal management in hot and dry climate region. Appl. Therm. Eng. **112**, 1240–1259 (2017). https://doi.org/10.1016/j.applthermaleng.2016.07.032
5. Khudhair, A.M., Farid, M.M.: A review on energy conservation in building applications with thermal storage by latent heat using phase change materials. Energy Convers. Manag. **45**, 263–275 (2004). https://doi.org/10.1016/S0196-8904(03)00131-6
6. Shilei, L., Neng, Z., Guohui, F.: Impact of phase change wall room on indoor thermal environment in winter. Energy Build. **38**, 18–24 (2006). https://doi.org/10.1016/j.enbuild.2005.02.007
7. Jamekhorshid, A., Sadrameli, S.M., Barzin, R., Farid, M.M.: Composite of wood-plastic and micro-encapsulated phase change material (MEPCM) used for thermal energy storage. Appl. Therm. Eng. **112**, 82–88 (2017). https://doi.org/10.1016/j.applthermaleng.2016.10.037
8. Meng, E., Yu, H., Zhou, B.: Study of the thermal behavior of the composite phase change material (PCM) room in summer and winter. Appl. Therm. Eng. **126**, 212–225 (2017). https://doi.org/10.1016/j.applthermaleng.2017.07.110
9. El Omari, K., Le Guer, Y., Bruel, P.: Analysis of micro-dispersed PCM-composite boards behavior in a building's wall for different seasons. J Build Eng **7**, 361–371 (2016). https://doi.org/10.1016/j.jobe.2016.07.013
10. Ling, H., Chen, C., Wei, S., et al.: Effect of phase change materials on indoor thermal environment under different weather conditions and over a long time. Appl. Energy **140**, 329–337 (2015). https://doi.org/10.1016/j.apenergy.2014.11.078
11. Zhu, N., Liu, F., Liu, P., et al.: Energy saving potential of a novel phase change material wallboard in typical climate regions of China. Energy Build. **128**, 360–369 (2016). https://doi.org/10.1016/j.enbuild.2016.06.093
12. Evola, G., Marletta, L., Sicurella, F.: A methodology for investigating the effectiveness of PCM wallboards for summer thermal comfort in buildings. Build. Environ. **59**, 517–527 (2013). https://doi.org/10.1016/j.buildenv.2012.09.021
13. Kong, X., Lu, S., Huang, J., et al.: Experimental research on the use of phase change materials in perforated brick rooms for cooling storage. Energy Build. **62**, 597–604 (2013). https://doi.org/10.1016/j.enbuild.2013.03.048
14. Kant, K., Shukla, A., Sharma, A.: Heat transfer studies of building brick containing phase change materials. Sol. Energy **155**, 1233–1242 (2017). https://doi.org/10.1016/j.solener.2017.07.072
15. Souci, O.Y., Houat, S.: Numerical study of thermos physical properties of a hollow brick filled by the PCM. J Mater Environ Sci **8**, 2213–2220 (2017)
16. Principi, P., Fioretti, R.: Thermal analysis of the application of PCM and low emissivity coating in hollow bricks. Energy Build. **51**, 131–142 (2012). https://doi.org/10.1016/j.enbuild.2012.04.022
17. Zhang, C., Chen, Y., Wu, L., Shi, M.: Thermal response of brick wall filled with phase change materials (PCM) under fluctuating outdoor temperatures. Energy Build. **43**, 3514–3520 (2011). https://doi.org/10.1016/j.enbuild.2011.09.028
18. Silva, T., Vicente, R., Soares, N., Ferreira, V.: Experimental testing and numerical modelling of masonry wall solution with PCM incorporation: a passive construction solution. Energy Build. **49**, 235–245 (2012). https://doi.org/10.1016/j.enbuild.2012.02.010

19. Li, L., Yu, H., Liu, R.: Research on composite-phase change materials (PCMs)-bricks in the west wall of room-scale cubicle: mid-season and summer day cases. Build. Environ. **123**, 494–503 (2017). https://doi.org/10.1016/j.buildenv.2017.07.019
20. Wang, X., Yu, H., Li, L., Zhao, M.: Research on temperature dependent effective thermal conductivity of composite-phase change materials (PCMs) wall based on steady-state method in a thermal chamber. Energy Build. **126**, 408–414 (2016). https://doi.org/10.1016/j.enbuild.2016.05.058
21. Wang, X., Yu, H., Li, L., Zhao, M.: Experimental assessment on a kind of composite wall incorporated with shape-stabilized phase change materials (SSPCMs). Energy Build. **128**, 567–574 (2016). https://doi.org/10.1016/j.enbuild.2016.07.031
22. Vicente, R., Silva, T.: Brick masonry walls with PCM macrocapsules: an experimental approach. Appl. Therm. Eng. **67**, 24–34 (2014). https://doi.org/10.1016/j.applthermaleng.2014.02.069
23. Wang, X., Yu, H., Li, L., Zhao, M.: Experimental assessment on the use of phase change materials (PCMs)-bricks in the exterior wall of a full-scale room. Energy Convers. Manag. **120**, 81–89 (2016). https://doi.org/10.1016/j.enconman.2016.04.065
24. Cunha, S., Aguiar, J., Ferreira, V.: Mortars with incorporation of phase change materials for thermal rehabilitation. Int. J. Archit. Herit **11**, 339–348 (2017). https://doi.org/10.1080/15583058.2016.1222464
25. Ramakrishnan, S., Wang, X., Sanjayan, J., Wilson, J.: Thermal energy storage enhancement of lightweight cement mortars with the application of phase change materials. Procedia. Eng. **180**, 1170–1177 (2017). https://doi.org/10.1016/j.proeng.2017.04.277
26. Theodoridou, M., Kyriakou, L., Ioannou, I.: PCM-enhanced Lime Plasters for Vernacular and Contemporary Architecture. Energy Procedia **97**, 539–545 (2016). https://doi.org/10.1016/j.egypro.2016.10.070
27. Kheradmand, M., Azenha, M., de Aguiar, J.L.B., Castro-Gomes, J.: Experimental and numerical studies of hybrid PCM embedded in plastering mortar for enhanced thermal behaviour of buildings. Energy **94**, 250–261 (2016). https://doi.org/10.1016/j.energy.2015.10.131
28. Lachheb, M., Younsi, Z., Naji, H., et al.: Thermal behavior of a hybrid PCM/plaster: a numerical and experimental investigation. Appl. Therm. Eng. **111**, 49–59 (2017). https://doi.org/10.1016/j.applthermaleng.2016.09.083
29. Sarı, A., Bicer, A., Karaipekli, A., Al-Sulaiman, F.A.: Preparation, characterization and thermal regulation performance of cement based-composite phase change material. Sol. Energy Mater. Sol. Cells **174**, 523–529 (2018). https://doi.org/10.1016/j.solmat.2017.09.049
30. Zhang Y, Cui H, Tang W, et al.: Effect of summer ventilation on the thermal performance and energy efficiency of buildings utilizing phase change materials. Energies. **10** (2017). https://doi.org/10.3390/en10081214
31. Ramakrishnan, S., Wang, X., Sanjayan, J., Wilson, J.: Thermal performance of buildings integrated with phase change materials to reduce heat stress risks during extreme heatwave events. Appl. Energy **194**, 410–421 (2017). https://doi.org/10.1016/j.apenergy.2016.04.084

Author Index

M. Awang and M. H. Isa (eds.), *The Advances in Civil Engineering Materials*,
Lecture Notes in Civil Engineering 19, https://doi.org/10.1007/978-981-13-2511-3

MIX
Papier aus verantwortungsvollen Quellen
Paper from responsible sources
FSC® C105338

If you have any concerns about our products,
you can contact us on
ProductSafety@springernature.com

In case Publisher is established outside the EU,
the EU authorized representative is:
Springer Nature Customer Service Center GmbH
Europaplatz 3, 69115 Heidelberg, Germany

Printed by Libri Plureos GmbH
in Hamburg, Germany